Do Animals Think?

Do Animals Think?

Clive D. L. Wynne

PRINCETON UNIVERSITY PRESS

PRINCETON AND OXFORD

Published by Princeton University Press, 41 William Street, Princeton, New Jersey 08540
In the United Kingdom: Princeton University Press, 3 Market Place, Woodstock,
Oxfordshire OX20 1SY

Library of Congress Cataloging-in-Publication Data
Wynne, Clive D. L.
Do animals think? / Clive D.L. Wynne.
p. cm.
Includes bibliographical references (p.).
ISBN 0-691-11311-4 (CL : alk. paper)
1. Animal intelligence. 2. Consciousness in animals. I. Title.
QL785.W9525 2004
591.5'13—dc22 2003060019
British Library Cataloging-in-Publication Data is available
This book has been composed in Sabon text with Pepita & Cochin Family Display
Printed on acid-free paper. ∞
www.pupress.princeton.edu
Printed in the United States of America
1 3 5 7 9 10 8 6 4 2

For Ros in all her species

Contents

1

What Are Animals?

I grew up on the Isle of Wight. "That southern island / Where the wild Tennyson became a fossil," W. H. Auden called it. A century ago it was very popular with poets. Today it is just a quaint English seaside resort: silent in the winter, bustling with tourists in the summer.

In 1994, long after I'd left on my travels, one Barry Horne woke up the sleepy Isle of Wight. He planted a succession of bombs. One in a pharmacy, another in a car parts store, a third in a fishing tackle supplier, and a fourth in a Cancer Research charity shop. The largest bomb caused nearly £3 million (about $5 million) of damage to the pharmacy. Horne chose that store because the parent company (Britain's largest pharmacy chain, Boots The Chemist) had in his view reneged on a promise to stop selling products tested on animals. The car parts store belonged to the same group as the pharmacy, so it had to go too. Fishing tackle—Horne was presumably objecting to angling. Cancer Research shops specialize in used clothing and household goods, sold by volunteers to raise money for research into cures for cancer. Where could be the objection to such good works? Cancer research involves animal experimentation, so the charity shop was a legitimate target also. He hid a bomb in a packet of cigarettes, which he stuffed into a leather bag. A Mrs. Woods bought the bag and took it away with her on a trip. Not knowing she was carrying a bomb, she let her children, aged three and six, play with her bag's contents. They found the fake pack of cigarettes. Fortunately, the device failed to explode.

I'm very puzzled by Horne. One thing I'd like to know is why he chose the Isle of Wight for his attacks. He was from Northampton, many miles away. Perhaps he had memories of a particularly miserable summer holiday on the beaches of the Isle of Wight. I resent that he chose to target a place so special to me.

But clearly for those of you not brought up on the Isle of Wight, the question is, Why would he plant bombs anywhere at all? Two things are worth noting here. The first is that there can be no doubting Horne's sincerity. In November 2001 he died of liver failure as a consequence of a hunger strike while serving an eighteen-year sentence for arson. He was trying to pressure the British government to set up an enquiry into animal experimentation.

The second important point is that Horne was not the British equivalent of the Unabomber (Theodore Kaczynski, a loner in Montana who directed bombs primarily at university personnel out of a deluded grudge against society). Horne was a member of the Animal Liberation Front—a group which refers to him on their web site as a "courageous fighter" with "thousands of supporters." It is difficult to gauge the degree of support for animal rights groups in the general population with any accuracy, but it is certainly not negligible: Horne's funeral was attended by three hundred people. So, though Horne obviously represents an extreme case, violent protest against what we do to animals is not limited to just one maverick.

I'm not trying to claim that all animal rights activists are vicious brutes, or that all the uses to which we put animals are defensible. What I am interested in is the range of attitudes in our society toward animals. Why, for example, given that Horne felt driven to wreak his vengeance on the Isle of Wight, did he choose to focus on shops? If he had to attack someone, why not the farmers on the hills? They do a lot more to animals on a day-to-day basis than the shopkeepers in the valleys. If Horne was trying to convert us all to a way of life less likely to harm animals, wouldn't we be more readily convinced to give up eating meat than medical treatment? Meat eating is nothing but an indulgence. Our animal-tested medications, on the other hand, are of proven effectiveness.

Even Linda McCartney, next to Mahatma Gandhi the most famous vegetarian and animal lover of the twentieth century, was fed anticancer drugs as she lay dying—drugs that had been tested on animals. News reports on this are conflicting: either her family did not themselves know that these drugs had been tested on animals, or they knew but were so desperate to see her get better that they did not tell her.

Barry Horne clearly thought he knew what animals were. He was so completely convinced that animals are sentient and worthy of protection that he was willing to lay down his life for them. Linda McCartney clearly also suffered little ambivalence in her attitude toward animals.

Sometimes I wish I could share their certainty.

What are animals—really? What should we make of them? Are they machines: complicated, intricate, beautiful perhaps, but fundamentally mechanical? Or are they something else; something conscious and thoughtful? Are animals possessed of some special spark that sets them off from the mechanical and vegetable worlds? Could they be both at once—conscious machines? Human beings have always lived among other species, and we fret, now perhaps more than ever, over the correct way to deal with them. How can we treat animals appropriately if we don't even know what they are?

As I grew up on the Isle of Wight, the relationship I and my family and friends had with animals expressed the same ambivalence to be found anywhere in the industrial world. We loved our pets, and at some level we knew that the meat on our table had come from animals that lived on the hills around us and that were not dressed up in bows for their birthdays, invited to sleep on their owners' beds, or talked to like bona fide human beings. We lived this dichotomy but, as far as I remember, were little troubled by it.

In those days the butchers' shops still had sawdust on the floor and cuts of meat in the window. The meat wasn't packaged on little plastic trays but cut from recognizable limbs of cattle, pigs, and sheep. And yet I don't ever remember being shocked by this. It was just the way of the world.

James Serpell of the University of Pennsylvania expresses the contradictions in our attitudes toward animals very clearly: "At one extreme are the animals we call pets. They make little or no practical or economic contribution to human society, yet we nurture and care for them like our own kith and kin, and display outrage and disgust when they are subjected to ill-treatment. At the other extreme we have animals like the pig on which a major section of our economy depends, supremely useful animals in every respect. . . . We pickle its trotters, make black puddings from its blood, sausages from its intestines, and expensive and durable leathergoods from its skin. We even emulsify its thick white fat for the production of ersatz ice-cream. And in return for this outstanding contribution we treat pigs like worthless objects devoid of feelings and sensations."

Can pigs feel? Are they capable of sensations? How would we know? And why doesn't anyone seem to care one way or another? The great mass of pig-eating humanity may not be bothered about the emotional and sensory worlds of the beast that ended up as bacon on the breakfast table, but to me more surprisingly, the animal liberation movement does not seem greatly interested either: The North American Press Office of the Animal Liberation Front issues an annual report on direct action by any individual or organization to liberate animals in the United States or Canada. The 2001 edition—dedicated to Barry Horne—documents the rescue of five thousand animals, not one of them a pig. Why not?

Many people are fascinated and intrigued by animals, and yet very few seem to be aware of the work that has been done in the last fifty years to improve our understanding of animal minds. True, there are a couple of high-profile "discoveries" that everybody knows about. If I had a penny for every time I have been told that chimpanzees are genetically as nearly identical to us as makes no difference and, given appropriate training, can communicate in human language, I would have a great pile of small change. My pockets would also drag if I collected coinage each time I am told that dolphins use an elaborate language among themselves that we are not smart enough to decode. But aside from these high-profile (and highly questionable) discoveries, the public at large just draws a blank. This book is my attempt to fill in some of that blankness, to bring some of the discoveries of modern animal behavior science to a wider audience.

THE SIMILARITY SANDWICH

What I want to do in this book is sweep all the debris of traditional views of animals, now mixed up with mauled science, right off the table and start again—that is, start with the reliable knowledge we have of what animals do. I have often been disappointed by how little scientific work is done on the psychology of animals. Too many psychologists define their sphere of interest as exclusively their own species, just *Homo sapiens*: How limited! So many exciting things have been discovered about all kinds of species. Just yesterday I was reading about the bolas spider that lures moths by imitating moth pheromones. In a transspecies war of the sexes, the female bolas spider imitates the attractant pheromone of the female moth, so only male moths are attracted to her. The spider has no web: when the moth gets close enough, she hurls a sticky ball (bolas) of silk at the moth to capture him. In America, bolas spiders just swing the bolas back and then out: in Africa and Australia these spiders swirl their bolases around like a lasso before lashing out at the moth. Either way, the moth gets caught, paralyzed, and wrapped in silk.

There is enough science out there to lay the foundation for an objective understanding of animals. I am convinced that we are beginning to know what animals are. And I can tell you up front: animals are not like us. But in many respects they are like us.

Other species are not like us. But they are also like us. I see a "not like us—like us—not like us" sandwich. It looks like this:

The bottom layer is a layer of dissimilarity. Each species on this planet lives in a unique sensory world. The sonar of the hunting bat (and the moth, its prey). The ultraviolet light seen by birds; the infrared of insects. The rich sense of smell dogs enjoy. The electric and magnetic fields to which some fish and a few other animals are sensitive. (The obscure Australian duck-billed platypus can tell if a battery has any current left in it—though there are easier methods of testing batteries.) Birds pick up on changes in air pressure. At this level there is no denying the diversity in the animal kingdom.

The middle layer is a layer of similarity. Here we find basic psychological processes like learning and some kinds of memory, along with simple forms of concept formation, such as identifying objects as being the same or different from other objects and a basic sense of time and number. All of these seem be common to a wide range of species and to operate in similar ways in animals as diverse as chicks and chimpanzees, spiders and squirrel monkeys.

It is not always easy to be sure whether something belongs in the middle layer (similarity) or the bottom layer (dissimilarity) of the sandwich. Studies on reasoning and problem solving, for example, do not seem to reveal many differences between species. But then again, so few species have been studied. And tool use (one of the few activities suggestive of reasoning that can be observed in the field) is definitely very different from species to species (and absent in most). Maybe it isn't a ham sandwich, but one filled with something that can squish into the bread—something like cream cheese? The boundary between the first two levels is not a firm one. My point is that every animal's world is different, just as every animal's lifestyle and niche are different. And yet there are also commonalities in animal minds, because we are all living on the same planet and descended from the same slimy ancestors.

But when we come to the bread on top of the sandwich, we notice something very different. After forty years of trying we can say definitively that no nonhuman primate (or any other species) has ever developed anything equivalent to human language. The hens in the chook house will never "hatch an elaborate plot to escape from the clutches of the menacing Mrs. Tweedy" (to quote from the web site for the movie *Chicken Run*). Most nonhuman species show very little interest even in imitating each other—let alone communicating with each other and coordinating their activity. Even chimpanzees, though they may recognize themselves in mirrors, are very slow to understand the motives of other individuals. They seem no better able to place themselves imaginatively into the shoes (or paws or hoofs) of another individual than are autistic children. This is a very surprising fact, and one that animal behavior scientists have been reluctant to face up to. There really is a difference between humans and other animals. A pretty big difference. The psychological abilities that make human culture possible—enthusiasm to imitate others, language, and the ability to place oneself imaginatively into another's perspective on events—are almost entirely lacking in any other species. They didn't have to be: we *are* all related, and we *do* share a great many psychological qualities with other species. But, as it turns out, we really are different from them in crucial respects.

DARWINIAN ACID

Acknowledging that the similarity sandwich has bread (that is, dissimilarity) on top, that there is something different about us humans, is a hard-wrung admission for me. I am a hard-core Darwinian. I believe that all species on this planet are related, some more closely than others, and that our common stock diverges through a process of selection. Some of us survive and thrive, have children and grandchildren. Others are left on the slag heap of (evolutionary) history. So long as children somewhat resemble their parents, natural selection will shape us to fit our environment.

In a wonderful exploration of Darwin's theory, Daniel Dennett, philosopher at Tufts University, called natural selection *universal acid*. He meant that it is an idea so powerful that nothing can contain it: it eats through every barrier. I like Dennett's metaphor because it emphasizes how difficult—even dangerous—it can be to work with very powerful ideas.

For example: We share 98.4 percent of our DNA with chimpanzees, and probably even more with bonobos (also known as pygmy chimpanzees). Does this mean that chimpanzees and perhaps other great apes share human self-awareness and are entitled to similar protection under the law as we are? This is what a group calling itself The Great Ape Project claims. The Great Ape Project includes among its supporters such luminaries as Oxford biologist Richard Dawkins (author of *The Selfish Gene*), Peter Singer (Princeton ethicist and author of *Animal Liberation*, one of the founding documents of the modern animal protection movement), animal rights lawyer and author of *Rattling the Cage* Steven Wise, and leading chimpologist Jane Goodall, who has studied chimpanzees in the wild for over forty years.

Another example: The notion of evolutionary continuity does not just apply to our closest relatives, the great apes; all animals on this planet are our relatives to a greater or lesser degree. Does this mean that they are all conscious (as the Harvard zoologist Donald Griffin has claimed in *Animal Minds: Beyond Cognition to Consciousness*), or at least that they all think (as Harvard psychologist Marc Hauser argues in *Wild Minds*)?

To me these examples show the difficulties of working with Darwin's powerful acid. Darwin's theory says we are all related, not that we are all identical. Every species has its unique adaptations. To those who would say that the human mind is a unique adaptation, I would say, "Balderdash." it bears some similarities to the adaptations of other animals, and presently I'll show them to you. But to those who would say that evolution dictates that there can be nothing unique about the things we humans do (and especially what we say to each other) I would say "Balderdash" again—or possibly something stronger. I can show you differences.

Darwin's theory also says that there are no magic sparks. No divine intervention separates us humans from all the rest of creation. In denying human-style language to any other species, I am not sneaking back in some special vital spark in the human case, I am not trying to lift humans up from the beasts and closer to God. Animal lovers have hated René Descartes for centuries for suggesting that animals are machines. And some would love Darwin for allowing that animals, through their relatedness to humans, could again share that special human spark—the soul that fires the mind.

But this is to read Darwin backward: Darwin's achievement is to let us see that we are all machines, mankind included. "Man . . . with divine face, turned towards heaven, . . . he is no exception," Darwin wrote. We are all machines: sea anemones; fish, dolphins, horses, golden retrievers, and bank managers. We are all machines designed by natural selection to solve the problems we confront in our daily lives to such a degree that we find the time to raise healthy, viable offspring who are likely to have healthy children of their own. Our minds and behavior are as much a part of the package of adaptations that sees us through life as are our anterior appendages and our feeding habits. And just as our anterior appendages and feeding habits show points of similarity and dissimilarity due to our shared (and not-so-shared) evolutionary histories and present-day environments, so too our minds and behavior are both similar and dissimilar. Counterparts to the bones in the human hand can be seen in the flippers of dolphins and the wings of bats. In just the same manner, some points of similarity can be seen between the minds of humans, dolphins, and bats. But do dolphins have "hands"? Do bats have flippers, or people wings?

To admit that humans are different does not return them to the center of the universe. I believe that you, dear reader, are a member of the species *Homo sapiens*, and I base that on nothing more than the fact that you are reading this text. Just as you can make the same reliable guess about me just by virtue of the fact that I wrote it. This is not a trivial observation. It is a distinction of some power between human minds and other minds. But it does not make the human *better*. Language is a powerful adaptation, but it is not always a power for good. Barry Horne's capacity for linguistic thought played some part in his demise. If he had not been able to formulate thoughts in words and be influenced by the words of others, he probably would not have killed himself.

The jury is still out on the whole human experiment with language and material culture. It is certainly possible that we may, through our diabolical ingenuity, create conditions on this planet that make our further survival impossible—hardly a favorable outcome from an evolutionary perspective. So acknowledging that language is a uniquely human adaptation does not return human beings to the pinnacle of some *scala naturae* (the medieval scale of beings, from snails to angels, that still structures most people's view of the animal kingdom), nor does it return the sun to revolving around the earth, or reverse any other step in the gradual displacement of our egotistical selves from the center of the universe.

I think it is important to understand how similar and dissimilar other animals' minds are to our own. Our opinions on what it is like to be a chicken will likely influence our attitude to battery housing and sex between men and hens (a practice that Peter Singer has suggested may be no crueler than battery housing and industrialized slaughterhouses). If we believe that chimps are self-aware, then that will influence decisions about their suitability for use in research on hepatitis—a disease that affects half the world's (human) population. At the moment chimpanzees are the only medical research model for hepatitis.

There are many practical questions, from the appropriateness of eating animals to whether our cat should be allowed to go out and hunt at night, where an objective understanding of what animals really are is badly needed. But, important as these issues may be, they are not what motivate me to worry about this question. What I want to know is this: Are we human beings—*Homo sapiens:* knowing man—alone on this planet in our consciously thinking minds, or are we surrounded by knowers whose thoughts are just too alien for us to understand? To contemplate this question is to stand, not on the edge of an abyss, but on the cusp between two abysses. Either outcome would be astonishing. To know for sure that we had thinking companions on this planet would be an amazing discovery. I find that at least as stunning a possibility as the discovery of minds on other planets—let's find the other minds on our own planet first! On the other hand, to know with the certainty that science can bring that we stand unique in our reflective, thoughtful intelligence—that would also give me to pause. I'd probably have to take the dog for a walk to absorb that one.

The New York University philosopher Thomas Nagel famously wondered what, if anything, it might be like to be a bat, "seeing" a world by listening to ultrasonic echoes. His conclusion was that we could never know. But scientists happily blunder in where philosophers fear to tread. There must be something we can know about the world of the big brown bat as it perceives in total darkness unevenness in a surface of just over one-tenth of an inch. Just as there must be something we can discover about the world of a noctuid moth—the prey of the bat. When the moth hears the bat's ultrasound switch to an attack pattern, it generates ultrasound of its own to jam the bat's sonar system.

The perceptual worlds of many species are so different from our own that it is perhaps not surprising that we have difficulty making sense of them. Rupert Sheldrake, in his popular *Dogs That Know When Their Owners Are Coming Home and Other Unexplained Powers of Animals*, is so baffled by the things that animals do that he believes we have to call in the supernatural to make sense of them. I am not willing to do that. There is nothing that animals do that will be made simpler by giving up on rational explanation.

In four of the chapters that follow I take one species (or group of species) and explore the world from that animal's perspective. I put on the skin of an insect (the honeybee) in chapter 2, a bird (the pigeon) in chapter 4, a flying mammal (the bat) in chapter 6, and a swimming mammal (the dolphin) in chapter 8. I selected these four species because their worlds were excitingly alien, but also sufficiently studied so that there is plenty to be said about them.

In the remaining chapters I tackle the three major faculties that have long been seen as discriminating humans from all other species: reasoning (chapter 3), language (chapter 5), and the ability to put oneself imaginatively into the position of another—what we could call "theory of mind" (chapter 7).

As a teenager walking our dog on the beaches of the Isle of Wight, I honestly used to think that nobody understood me better than that animal. The image of the cliffs, the beach, and the dog in winter is so deep in me it's intoxicating just to think about them. Benji has long since passed on to doggy heaven, but I still rate the companionship of animals as one of life's highest joys. And I agree with James Serpell that just because pet keeping is sentimental doesn't make it a bad thing: as Serpell says, many of life's most rewarding moments are beset with sentimentality. But I do now strive for an objective understanding of animals. And that objectivity tells me that Benji's understanding of me was as dim and restricted to patterns of comprehension appropriate for his species as mine of him was constrained by my human thinking. That doesn't make Benji and his kind (and all the other kinds) less interesting—only more. What I want now is to get to what animals really are, not the sentimental version of what they seem to be. This isn't as easy as talking to our pets and assuming they understand us. But it is, I think, ultimately more satisfying.

FURTHER READING

The Truth about Dogs, by Stephen Budiansky (Viking, 2000). Witty, intelligent, meditative, and shocking: this is one of my favorite books on any species.

In the Company of Animals: A Study of Human-Animal Relationships, by James Serpell (Cambridge University Press, 1996). This is a modern classic on the relationship between people and other species.

Darwin's Dangerous Idea, by Daniel Dennett (Touchstone Books, 1995). Darwin's theory of evolution has inspired many good books, of which this is one of the most thoughtful and interesting.

The Animal Estate, by Harriet Ritvo (Harvard University Press, 1987). Ritvo takes as her canvas England of the Victorian age and explores the evolving attitudes toward animals during that tempestuous time.

Animal Minds: Beyond Cognition to Consciousness, by Donald R. Griffin (University of Chicago Press, 2001). Griffin takes the extreme position that all animals are conscious.

Wild Minds, by Marc Hauser (Henry Holt, 2001). Hauser is less extreme than Griffin, arguing, not that all animals are conscious, but that they all think.

The Great Ape Project: Equality beyond Humanity, edited by Paola Cavalieri and Peter Singer (St. Martin's, 1994). This is perhaps more a political work than a strictly scientific one; the various authors plead for legal protection for the great apes.

2

The Secrets of the Honeybee Machine

My earliest recollection of bees—indeed, one of my earliest memories of any kind—is of being stung by one. I was putting on my shoes on the stone floor of the changing room at Shanklin C of E Primary school. ("C of E" stands for Church of England. They asked me where my family went to church. I said we didn't. They said, "Well then, you're C of E.") I put out my hand to steady myself and felt a sharp pain in my palm.

There is so much more to know about honeybees than just that they sting. Honeybees know the time of day; they know the points of the compass. Honeybees forage over a four-mile radius and find their way home to the hive and back to the productive flowers. Not only can an individual bee return to flowers it fed from earlier, but it can communicate to other bees where these flowers are to be found. In fact, honeybees have one of the most complex forms of communication in the animal kingdom. No other creature besides the human can convey three dimensions of experience to its fellows (honeybees communicate the distance, direction, and quality of a food source to their hive mates). Honeybees live in societies more rigidly structured than the army. Different bees fulfill different roles, from the queen, who lays all the eggs and controls the entire hive, to the little worker bee on air-conditioning detail, who fans fresh air around the hive with her wings. Honeybees scout out good sites for hives, communicate with each other about possible locations, reach a consensus, and then fly off to build an elaborate new home.

With all this ingenious behavior, particularly their complex communicative abilities, should we get mystical about honeybees—consider them conscious? When the pioneer Swiss psychoanalyst Carl Jung heard of the ability of honeybees to communicate he wrote: "This kind of message is no different in principle from information conveyed by a human being. In the latter case we would certainly regard such behavior as a conscious and intentional act and can hardly imagine how anyone could prove in a court of law that it had taken place unconsciously." This passage is quoted most approvingly by distinguished Harvard zoologist Donald Griffin, who, in his *Animal Minds: Beyond Cognition to Consciousness*, claims that bees are conscious. Griffin, however, fails to quote Jung's next sentence, "Nevertheless it would be possible to suppose that in bees the process is unconscious."

Griffin is in a vanishingly small minority of those with any interest in honeybees. For most of us, whatever our feelings about cats and dogs, and no matter how clever a bee may turn out to be, these insects are nothing but machines. And yet what wonderful machines! A close look at the bee machine will perhaps make the notion of all animals as machines more palatable.

Everything honeybees do they achieve with just a sand grain's worth of brain. In fact, bees don't really have brains at all—just a string of nerve fibers down their back, with collections of nerve cell bodies in clumps called ganglia (singular: ganglion). Each ganglion contains a few tens of thousands of nerve cells. At the head end, there is a brain of sorts formed by the fusion of a few ganglia. This brain contains fewer than one million nerve cells. One million may sound like a large number, but in computing terms it really isn't. Our human brains contain ten thousand times as many nerve cells. Even the computer on which I write contains at least ten times more active parts than a bee's brain. And yet the honeybee achieves so much with so little. How can that be? What are the secrets of the honeybee machine?

Let's start with a little honeybee history. Honey is the sweetest and most energy-rich naturally occurring food known to mankind, so it's not surprising to find that people have been stealing honey from honeybees for countless millennia. At first just by raiding wild bee hives—cave paintings in Spain dating back around eight thousand years show humans hunting honey. One thousand years later they began to build artificial hives. Pottery vessels and wicker baskets for holding bees are known from the Neolithic period. Hollowed out logs have also been used as beehives since about 2000 B.C. Wall paintings from ancient Egyptian tombs show honey being harvested from artificial hives around 1450 B.C. The modern hive, with its easily removed racks of honey-filled comb, was invented in 1851 by a young pastor from Andover, Massachusetts: Lorenzo Lorraine Langstroth. Today, around 50 million Langstroth hives, situated all over the world, are producing over 2 billion pounds of honey a year.

Honey is one of only two animal bodily secretions that people eat without killing the animal: The other is milk. But bees were holy before God promised the Jews a land of milk and honey. Zeus, the king of the gods in Greek mythology, was protected and fed honey by sacred bees. Apollo and Dionysus were also fed honey by bee maidens (that's what nymphs were). And many other mythic figures in ancient tales were served by bees and had bees associated with their worship.

The dominant honey-producing bee in the world today, *Apis mellifera*, originated in Europe but has been spread by human intervention throughout the world. People in different places at different times have exploited a number of other species of bee, but these have largely been forgotten as *Apis mellifera*, with its greater honey producing potential, has supplanted them. As well as the nutritional benefits from honey, honeybees also perform a useful job in pollinating crops as they forage for nectar from flower to flower. It is because of their economic importance that we know so much about bees. What we know about honeybees must stand proxy for what might be knowable about other insects, if only we invested more effort in finding out about them.

MEET THE FAMILY

The biggest clue to the amazing abilities of honeybees, in particular their communicative skills and cooperative activity, comes from understanding the structure of the family that calls the hive home: one mother; tens of thousands of offspring. If you think your relatives are a little strange, wait till you see how things look at the honeybee family gathering.

There are three castes of honeybee. Any one beehive usually contains only one queen. She is the fertile female, the mother of all the others. Of fathers there are none. A few hundred, or perhaps a thousand, drones (who are tolerated only in the spring and summer) are the token males. Then there are thousands or tens of thousands of workers—the drudges (all female) who do all the work, from feeding the larvae, through foraging for pollen and nectar, to defending the hive from invaders. There are even workers that keep the hive clean.

Starting at the top: the queen is the absolute ruler of her hive. In the normal run of things she is the only individual who lays eggs: all the drones and workers are her offspring. She controls the activities of all her subjects with powerful chemicals called pheromones which they have no choice but to obey. Except for a couple of days at the beginning of her life, when she takes a number of nuptial flights and mates on the wing with about a dozen male drones from other hives, the queen spends her entire life in her hive. Each year of her one to three years of life she lays around two hundred thousand eggs.

Charles Butler published *The Feminine Monarchie* in 1609, three years after the death of Queen Elisabeth I of England. But his was not an account of the life and times of "Good Queen Bess"—not the first female monarch of England, but surely the most powerful and impressive. Rather, Butler's work is a surprisingly accurate "treatise concerning bees, and the due ordering of them."

Butler read in the orderly lives of honeybees a God-given lesson of the "most natural and absolute forme of government . . . *Monarchie*"; furthermore, he noted that "The males heer beare no sway at al, this being an *Amazonian* or *feminine kingdome*."

Though he did not know how she achieved it, the total power of the queen over her subjects was clear to Butler: "If by hir voice she bid them goe, they swarme; if being abroad she dislike the weather, or lighting place, they quickly returne home againe; while she cheereth them to battaile they fight, when she is silent they cease, while she is well, they are cheerefull about their worke, if she drope, they faint also; if she dy, they will never after prosper, but thenceforth languish till they bee dead too."

Queen Elisabeth I ordered the execution of her cousin, Mary, Queen of Scots, when she believed she had become a dangerous rival to the throne. Queen bees are not so different. In the formation of a new hive, several cells containing larvae with the potential to become queens are laid down. The first queen to emerge from her cell in the early days of a new hive destroys the other queen pupae. With her powerful mandibles she bites a hole in the side of each of the other queen cells and then inserts her abdomen and stings her rival to death. Workers dispose of the carcass and destroy the empty cell.

Drones are the males in the hive. They do no work; their only function is to mate with the queens of other hives. On balmy summer afternoons they fly out looking for a queen for sex. If this sounds like the best job in bee-land, keep in mind three things. First, most drones die without ever mating (there are many thousands more drones than queens). Second, a drone that is successful in finding a queen and mating with her dies immediately as his abdomen bursts during copulation. Third, those drones that survive a whole summer of failed sexual kamikaze missions are kicked out of the hive by their worker sisters and freeze to death as the hive prepares for winter. The average drone life span is three to four weeks. So life as a drone is not so breezy after all.

The female bees do all the work. They build, defend, ventilate, repair, and clean the hive, tend to the queen and her brood, and forage for and store pollen and nectar. The workers are all sisters.

and the hiuing of them. C,5

for a ſwarme: which ſeldome ariſeth the next day, vnleſſe the weather be very pleaſãt: but af-ter two or three daies they will accept indiffe-rent weather. I haue not knowne any ſtay after the fift day.

24 *The Bees muſicke.*

They ſing both in triple time: the princeſſ thus

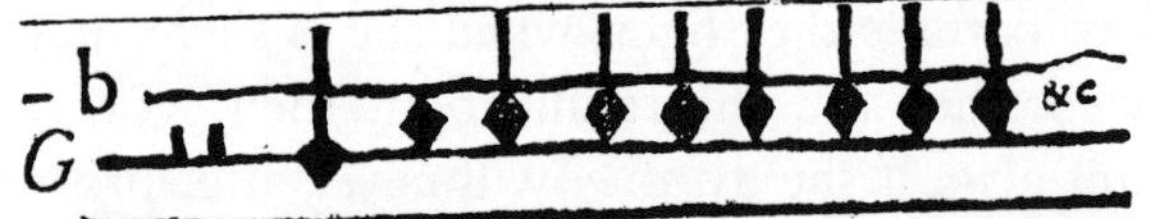

with more or fewer notes, as ſhe pleaſeth. And ſometime ſhe taketh a higher key, ſpecially to-ward their comming forth, and beginning the od minim in *A la mi re* ſhee tuneth the reſt of hir notes in *C ſol fa* thus,

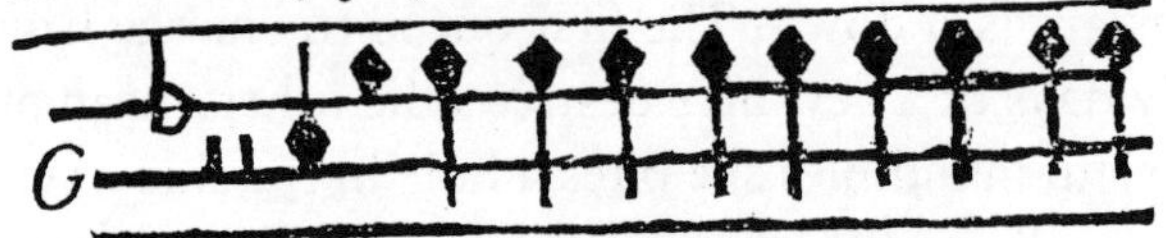

But the Queene in a deeper voice thus,

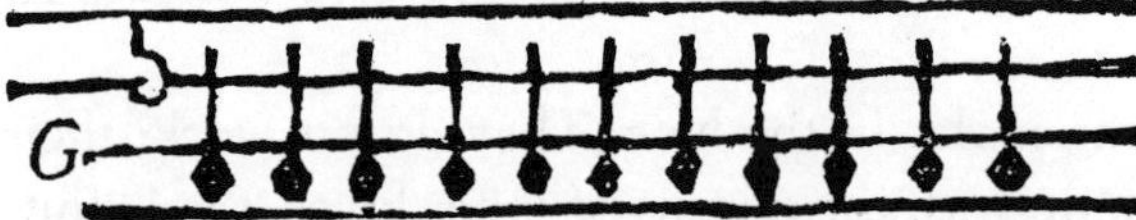

continuing the ſame, ſome foure or fiue ſemi-briefes, and ſounding the end of every note in *C ſol fa vt*. So that when they ſing together, ſometime they agree in a *perfect third*, ſome-time in a *Diapente*, & (if you reſpect the termi-nation of the baſe) ſomtime in a *Diapaſo*. With theſe tunes anſwering one another, and ſome

F pauſes.

The various songs of the honeybee as notated by Charles Butler in his 1609 treatise, *The Feminine Monarchie*.

Contrary to the stereotype of the busy bee, however, researchers have noted that worker bees spend most of their time resting and patrolling the hive. They also sleep at night. Workers live around a month in the summer and up to six months in the winter.

All three types of honeybee hatch from eggs laid by the same queen—so what makes the difference between queen, drone, and worker? A queen, having mated with drones from other colonies, takes their sperm back to the hive with her but does not release sperm into all the eggs she lays. An egg that is going to develop into a drone receives no sperm; it is unfertilized, and therefore all its genetic material comes from its mother. Consequently, a drone is genetically identical to the queen that laid it. Eggs destined to be queens and workers are fertilized with sperm. What makes the difference between developing into a mighty queen or a lowly worker is the quality of diet the developing larvae receive from the nurses (one specialized kind of worker bee) who tend the brood. Special food, a rich saliva known as royal jelly, is fed to eggs that become queens. No royal jelly and the female becomes a sterile spinster worker who will spend her life tending someone else's offspring.

In an ordinary human family (or the family of any other mammal, for that matter), each child receives half its genetic material from each parent. Each fertilization of egg by sperm is a new throw of the genetic dice: a new mixture of equal shares of paternal and maternal essences. Consequently, we each share, on average, half our genes with our father and half with our mother. Our (full) brothers and sisters were created the same way, and so they also share, on average, half their genetic material with each of their (our) parents and also share half with us. So a human family shows a symmetrical pattern of relatedness: each parent is equally related to all of his or her children, and the children are just as related to each other as to their parents. Odd exceptions do crop up, such as identical (monozygotic, as they are more technically known) twins. Monozygotic twins are the product of a single egg and sperm and so have only one set of genetic information between them: they are effectively clones of each other. Apart from twins, the only other way that the normal mammalian pattern of symmetrical inheritance can be upset is through incest. If a father were to have a child by his own daughter, he and this child/grandchild would share three-quarters of their genes rather than half, as is normally the case. In many, though not all, human cultures, such behavior would be considered immoral, illegal, or both. Among honeybees, however, these more complicated patterns of inheritance are perfectly normal.

Let's start with the drones—the tiny proportion of honeybees who are male. Drones, as already mentioned, are produced from the queen's unfertilized eggs. They are consequently genetic clones of the queen. The drones go off and try to fertilize virgin queens from other hives. On a mating flight a queen accepts up to sixteen drones. Each drone dies as he copulates: perhaps that's why the French call orgasm *la petite mort*. With her afternoon in the sun completed, the queen carries the sperm of all the males she has mated with back to her hive to use during her career of egg laying. She may lay up to two thousand eggs each day, and to most of these she adds sperm. For reasons to do with the complicated genetics of the honeybee, each of these fertilized eggs turns into a female. Only one in many tens of thousands of these females will become a queen herself: most become workers. But whereas, in human and other mammal societies, sisters share half their genes with each other and half with each parent, by some tricks of bee genetics—having to do with the fact that bee fathers (drones) have only one set of genetic materials while bee mothers (queens) have two (geneticists call this haplodiploidy)—worker sisters of the same father share three-quarters of their genetic material with each other: these are called "super sisters." Consequently, these sisters of the same father are more closely related to each other than they are to their own mother. They are also more closely related to each other than they would be to their own daughters. But the queen does not usually mate with just one drone: she brings back to the hive the sperm of about a dozen different drones, each of whom fathers a distinctive "patriline." Daughters in different patrilines are only half-sisters; they share only one-quarter of their genetic material. This is because bee fathers only contribute half as much genetic material as bee mothers.

Why should we care about these funny patterns of genetic relatedness? Does it make any difference to honeybee family life that sons are clones of their mothers, and daughters are more closely related to their supersisters than to their own mother or children (if they have any)?

The genetic relations of the honeybee are important because they set the stage for honeybee social life. Honeybees are not just social animals; they are *perfectly* social. They do not just help each other out a little along the way; they dedicate their lives to each other. Worker bees sublimate almost all hope of having offspring themselves to helping their mother produce more young. Drones die for one chance at sexual intercourse. The specialties of bee genetics set the stage for this tight-knit social existence and the conditions that support honeybee communication.

The pioneer British geneticist J.B.S. Haldane is said to have quipped, "I would give my life for three brothers or nine cousins!" What Haldane was getting at was that a person is just a gene's way of making another person—a point Oxford zoologist Richard Dawkins has hammered home in more recent years with his concept of the "selfish gene." Since we share one-half of our genetic material with our siblings, and one-eighth with our cousins, the survival of anything more than two brothers or eight cousins should please us as much as our own. Though this is hardly a complete explanation of human altruism, patterns of human generosity do indeed show very preferential treatment of close kin. Rare is the (living) individual who donates a kidney even to a close friend; such a sacrifice is usually reserved for a family member. Family members tend to be the major beneficiaries of a deceased person's estate. It is most often the childless who leave large sums to charities dedicated to helping strangers.

If we were more closely related to our siblings than just 50 percent, the genetic calculus of altruism would change accordingly. For worker bees, who share 75 percent of their genes with their emerging supersisters, dedicating oneself to caring for one's sibs is a very worthwhile enterprise. If Haldane would lay down his life for three brothers, a honeybee can happily give her life for just two supersisters. This pattern of ultraclose genetic relatedness is the key to understanding why honeybee society is so tightly knit and how a relatively complex form of communication could have evolved in a beast with so little brain.

Cooperation likely evolved in humans because with our large brains we are well able to keep track of who is doing favors for whom. With that information we can be efficient and selective in choosing whom we assist. You scratch my back and I'll consider scratching yours. Honeybees, with their tiny brains, don't have that kind of cognitive apparatus available to them, but because of their very close genetic relatedness the engine of natural selection pushes them more strongly to help each other. Genes that encourage honeybees to help their supersisters are highly likely to be selected and recur in future generations.

If it really is genetic relatedness that causes honeybees to be so cooperative, then we might expect workers to be more disposed to assist their supersisters, who share 75 percent of their genes, than their half-sisters, who share only 25 percent. Sure enough, recent experimental work has demonstrated that honeybees can recognize the characteristic smell of their supersisters and treat them preferentially. Forager bees are more likely to hand over the pollen and nectar they have collected to supersisters than to half-sisters. When foragers return from the nectar hunt, they perform a characteristic dance to indicate where they found this food source (of which more in a moment); the bees that attend to this dance and learn where the nectar comes from are more likely to be the forager's supersisters than her half-sisters. Introduced to unfamiliar sister bees, workers are more likely to bite their half-sisters than their supersisters. The workers' recognition of kin extends to queens too. When a new colony is preparing queen larvae, nursing workers are more likely to care for the queen larva to whom they are most closely related.

Most of our understanding of the importance of genetics in shaping the closely knit community of the hive has come about in the last twenty years—genetics has been the most dynamic biological science of the late twentieth century. But the origin of the modern understanding of the patterns of inheritance is customarily ascribed to the pioneering work of a mid-nineteenth-century Czech abbot, Gregor Mendel. Mendel's famous research was carried out on different strains of pea plant; he patiently crossed varieties with different characters and looked to see where these cropped up in the succeeding generations. Originally, however, Mendel had intended to study honeybees. Despite designing a special mating cage, the abbot was unable to get the bees to mate on command. Without control over their mating, there was no way he could study patterns of inheritance. Hindsight shows that it is just as well that Mendel could not make a start on bees: their patterns of inheritance are far too complex for a beginner to grasp. If Mendel had stuck with bees, he would never have found the simple patterns of inheritance that gave him his great insights into genetics. It would take a century for these insights to be applied, by other researchers, to the honeybees that he had wanted to study.

FORAGING AND FINDING HOME AGAIN

So it is the honeybees' complicated genetics that sets the stage for a highly cooperative family life in the hive. And it is this cooperative home life that primes the conditions for the most humanlike and conscious-seeming behavior of bees, their ability to communicate. It's time to take a closer look at the coordinated activities of the honeybee: their cooperative foraging, and in particular their communicative abilities.

If you live near a hive of honeybees, you can carry out a simple experiment. Place a shallow saucer of sugar water on a windowsill, preferably close to some flowers where you have previously seen bees hanging out. (Unless you want a house full of bees, it's best to make sure the window is closed.) Sit and watch quietly outside the window with a pot of nail polish at the ready. When the first honeybee arrives, wait till it is engrossed in filling up on sugar water before very carefully placing a dot of nail polish on its back. Bees don't usually sting when they are away from the hive so you should be safe, but if you're not confident that you can do this without harming the bee, just leave it—you can still observe interesting things without touching the bee. A little later (how much later depends on how far your windowsill is from the hive), some more bees will head for your water saucer. These new bees will appear before your original honeybee makes it back to you. If you are a dab hand with the nail polish, you may be able to label these bee-buddies in a different shade. If you don't use nail polish, you will just see that a larger and larger number of honeybees come to your sugar-water saucer.

These simple observations raise two questions. First, how did that first bee find its way back to the hive (which could be up to four miles away) and out to your windowsill for a second visit? Second, how did the other honeybees know about your sugar water?

Honeybees have a very well-equipped navigational toolbox. The first tool in the box is the sun compass. Well-trained Boy Scouts can use their watches as a compass; honeybees do something similar. Scouts find compass points by pointing the 12 on their watches at the sun: south is now halfway between the hour hand and the 12. Knowing south they can identify all the other points of the compass. This works because the sun progresses from east to west via south in the course of each day. So if we know the time and the position of the sun, we can find compass directions. It's a bit like using a sundial backward. Honeybees, like most animals, have an inherent sense of time of day and are able to combine this with the position of the sun to intuit compass directions.

But how do they cope on overcast days? Boy Scouts know to use the magnetic compass in the heel of their boots when the sun doesn't shine—what do bees do? Honeybees continue to use the sun compass, even when the sky is largely (though not completely) cloud-covered. Unlike many other species (humans, for example), bees are sensitive to the polarization of sunlight.

To understand polarization we must recognize that light is electromagnetic radiation that comes at us in waves of energy. Just as a clothesline can be plucked to oscillate vertically, horizontally, or at any angle in between, so too light waves oscillate in characteristic orientations. It is this orientation of the waves that is known as their polarization. If this all sounds rather abstract, then that is because we are unable to perceive it directly ourselves. The only time we become aware of light polarization is when donning polarized sunglasses. Light reflected off water and glass tends to be oriented (polarized) in just one direction. By uniquely blocking just this one orientation of light, polarized sunglasses are able to reduce the glare in reflections without dimming light coming from other sources.

Honeybees, like many insects and birds, are able to detect the polarization of sunlight directly. This polarization is visible to them in any patch of blue sky. Furthermore, the polarization of sunlight varies as the sun moves across the sky. Consequently, a beast like a honeybee that can perceive polarization can figure the sun's position from any patch of blue sky, even if the sun itself is behind clouds. The bee then combines this information about the sun's position with her sense of what time of day it is to find compass points. Polarization of light is particularly strong in ultraviolet light—and bees are sensitive to ultraviolet.

But a compass alone is not a complete navigational system. To find our way back to some place, we also need a sense of distance covered. The next item in the honeybee's toolkit is an awareness, known as "dead reckoning," of how far it has traveled. The bee knows the distance it has traveled, not through how much energy it has expended in flying, nor in the sensation of air moving past it, but on the basis of the visual image moving past its eyes. Mandyam Srinivasan and his colleagues at the Australian National University demonstrated this through a series of elegant experiments with bees in wind tunnels. To understand the ingenuity of these experiments, it is first necessary to know more about the most amazing skill of honeybees: how they communicate to each other the location of food sources they have uncovered.

If you used two colors of nail polish on your neighbor's honeybees, then you know that the original forager bee that first found your saucer of sweetened water was overtaken on her return trip by some bee-buddies who had not been with her on the first trip. How did this second wave of bees know about your windowsill?

It turns out that the most remarkable example of communication outside human language does not come from the chicken (as the makers of *Chicken Run*, with the support of the American Humane Society, would have us believe), nor does it come from a primate. In fact, it doesn't come from a bird or a mammal, but from an insect. In the Bavarian Alps, in the final summer of the Thousand Year Reich, Karl von Frisch made a series of observations on honeybees for which he shared in the only Nobel Prize ever awarded for research into animal behavior.

Von Frisch pursued his research amid the most terrible impediments. In January 1941, unable to prove that his maternal grandmother had not been a Jew, he was declared a "mongrel second class" and banned from the university. Fortunately for him, however, a major plague in the honeybee populations of the Reich convinced the authorities to postpone his redundancy until the end of the war. But, as von Frisch was reprieved from one side, his safety came under attack from the other. On July 12, 1944, von Frisch's home near Munich was destroyed in an Allied air raid. The next day Allied bombers severely damaged his research center. However, no one was harmed at his home, and all the research equipment had previously been moved out of Munich and into the Alps. There, surviving on home-grown vegetables, von Frisch and his coworkers carried on their work right through the final days of the war, until, around midday on June 15, 1945—five weeks after the German surrender—he made his famous discovery.

When a honeybee returns with nectar and pollen from a successful foraging trip, she performs a dance on the vertical surface of a honeycomb. If the source of her provisions was less than about one hundred meters away, she performs a simple circular dance known, for obvious reasons, as the round dance. This dance lasts about thirty seconds, during which time other worker bees in the hive get very excited and troop around behind the dancing bee in a sort of bee conga. The lead bee's energy and excitement impart to her followers a sense of how good the source of nectar is that she has found, and the new recruits also pick up, with their antennae, the odor of the nectar source on the dancer's posterior. They then use this information to fly off and find the right flowers themselves.

If the source the first honeybee has found is further away from the hive, then the returning bee performs a dance on the vertical surface of the hive that is in a fattened figure-eight shape. She first dances a short vertical stretch before turning alternately to the left and right. Each turn concludes with a circuit back to the vertical, so that a reclining figure eight is drawn out, something like this—∞.

According to von Frisch, the follower bees recruited by the figure-eight dance do not just pick up the odor of the nectar source and the enthusiasm of the dance leader with their antennae; they also derive the distance of the nectar source from the hive and the bearing toward the sun on which they need to fly to reach these desirable flowers. As the forager bee dances, she waggles her abdomen, and von Frisch found a correlation between the duration of each waggle in the vertical part of the dance and the distance of the source from the hive. Interestingly, different races of bees use different metrics: for the German bees von Frisch studied, one waggle in the dance indicates about fifty meters from the hive; for Italian bees, one waggle is twenty meters; and for Egyptians, each waggle is just ten meters. Furthermore, he noted that the central part of the dance was not always danced strictly vertically. The angle of the central part of the dance to the vertical was the angle to the sun on which a bee had to fly to return to that source of nectar.

The dance communication system of the honeybee is amazing in itself (and not a little controversial, a point I shall come back to a moment). For our present purposes what is wonderful about the bee's dance system is that it provided a way for Srinivasan and his colleagues at the Australian National University to ask the bees that they put in a wind tunnel how far they thought they had traveled.

First Srinivasan allowed a honeybee to fly through the tunnel with everything turned off: there was no wind, no funny markings on the walls—just a plain old tunnel. At the end of the tunnel the bee found some sugar water and, after taking her fill, flew back to the hive. Back in the hive, the forager bee danced to indicate to her hive mates how far she had traveled to collect this food. Srinivasan and his team videoed the dancing bee and were able to measure from her dance how far she believed she had traveled. They also allowed other bees to observe the dances and then measured how far these recruited bees flew in the world outside the wind tunnel.

Next the research team tried to trick the foraging bee by blowing winds over it. The foraging bee returned to the hive, and its dance was videotaped. Would the bee think that a journey that had been more tiring because of the strong head wind was actually a longer trip? No departure from the original condition could be found: the bee was not confused about the distance it had traveled. Both the human and the bee observers agreed that the bees that had been in the wind tunnel were quite unfazed by the different head and tail winds they experienced, even though these changed the sensation of air moving past them and the amount of energy required to fly down the tunnel and back again.

Next the researchers marked the walls of the tunnel with patterns that could move at different speeds, to create the illusion that a longer or shorter distance had been covered than was truly the case. Again the forager bee flew down the tunnel, found sugar water, returned to the hive, and danced for her hive mates, and again Srinivasan and his colleagues estimated from her dance how far she thought she had flown. This time they found clear evidence that the bee was confused. Of course, this makes good sense: the real world outside the laboratory is often windy, but the real-world background against which a forager bee flies will rarely speed up or slow down—so it is this visual backdrop that supplies a firm basis for estimating distances covered.

CAN HONEYBEES READ MAPS?

So foraging honeybees have the sun compass to find bearings, and they have dead reckoning based on the visual background to estimate distances. Is there anything else in this little forager's toolkit?

One reason to ask if honeybees have anything else available to them on their four-mile foraging trips is because dead reckoning is an inexact business. As people lost in featureless deserts soon notice, small errors at the beginning of a journey progressively accumulate as the trip continues, so that by the time one should be back home, it is easy to be hundreds of miles off course. For more accurate navigation, people rely on maps and landmarks. Do honeybees have any abilities like this? Research from several groups in Europe and North America has demonstrated unequivocally that honeybees have an ability to use landmarks that they see along their route. Exactly how they do this, however, is still the subject of some controversy.

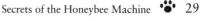

Foraging honeybees flying to and from the hive use landmarks, such as rows of trees, hills, or other salient features, to find their way. As they fly out, bees commit these scenes to memory and later use them to guide themselves on future trips. This has been demonstrated by experiments in which bees were captured and then released in the vicinity of familiar landmarks but out of sight of the hive or foraging site. Experiments with a movable landmark (a car) point to a similar conclusion.

James Gould, an expert in honeybee habits at Princeton University, claimed that bees use landmarks in a very complex way. Gould proposed that honeybees not only recognize how landmarks look when they are heading to or from a foraging site but integrate their representations of these landmarks into a proper map. A map is something much more than just a set of landmarks with information about which way you should head as you pass by them. We can call the use of landmarks without a map a route sketch: a set of directions relating to different landmarks. I can find my way to an unfamiliar spot if I am given information like "turn right at the lights by the vet," "head straight past the football field," and so on. This is a route sketch. If I get lost, however, or want to approach this unfamiliar spot from a different direction, then this route-sketch method is completely ineffective. If I have a proper map, on the other hand, I should be able to find my way even if I take a wrong turning, and I can find my way no matter which direction I am coming from. A map is a far superior instrument to a route sketch, but it is also far more complex. Is it really possible for the sand-grain-sized brain of the bee to contain a map?

The most direct way to test if a person or a honeybee really has a mental map or is just relying on landmarks is to first let them set off on their travels; then capture them, blindfold them (or, if they are bees, put them in a darkened container like a match box); and remove them to another place; then release them to see if they can figure out which way to head. This is just what Gould did with his bees, and he found that after being captured and displaced by the experimenters, the bees were able to correct their course for their goal when released. Consequently, Gould concluded that the bees were using mental maps. For several years, other researchers attempted to replicate Gould's results without success. Finally the explanation for these disparate results seems to have been found.

Imagine traveling in a car using a route sketch to find your way somewhere. If you are displaced from your route and have to find your way back, you will, in general, be completely lost. However, if your route sketch includes things like "head for the mountains," or "keep the Eiffel Tower on your left," you may well be able to find your way even after quite large displacements, because such major objects may still be visible. It seems that this was the situation for Gould's displaced bees. From their displaced positions they were still able to see some relevant landmarks.

To better test whether honeybees have mental maps, one must displace them to a position where landmarks are definitely not visible. A sometime collaborator of Gould's, Fred Dyer, now at Michigan State University, displaced bees while they were out foraging. He compared their behavior when they were displaced so that the original landmarks were still visible with their behavior when they were displaced so that they could no longer see any landmarks. It was only when the landmarks were still visible that the bees successfully continued in the direction of the feeder they had been heading for before the experimenters displaced them. Removed from the area where the familiar landmarks were visible, the bees were completely lost.

Thus, pending further studies, it seems that bees, though they remember landmarks and how to fly round them, do not possess a true map for navigation.

But maybe the map/no map dichotomy puts things too starkly. Foraging honeybees that are confronted by an obstacle (such as a tall building or mountain) that they must fly around to reach a food source can remember the detour they have to make. But what do they communicate to their hive-sisters when they get home—the true bearing of the food source from the hive, or the direction they actually flew out of the hive in order to avoid the obstacle?

Under these conditions, the foraging bee's dance indicates the true bearing of the food source—a compass bearing these bees have never actually flown but which they must somehow have deduced from the detour that they took. This, to me, indicates an integration of information about bearings and landmarks that is closer to what we mean by using a map.

Honeybees navigate by remembering landmarks and even integrate their knowledge of landmarks to some extent in a way similar to what we do with maps. Furthermore, bees communicate to each other about their foraging experiences. The second wave of honeybees that came up to your windowsill got there because the first forager bee told them where to go. What an amazing discovery that is—small wonder von Frisch received the Nobel Prize. But little wonder, too, that this discovery has attracted controversy. To concede such skills to a humble honeybee would seem to threaten our sense of the right hierarchical ordering of species.

Honeybees apparently communicate three dimensions of experience to each other: quality, distance, and sun bearing of a source of nectar. "Nectar of excellent quality can be found at 450 m on vector 090." No chimp, baboon, or vervet monkey can communicate that much to the rest of its troupe. This seems a little surprising, to say the least. In mitigation, we have already seen that hivemates are much more closely related to each other than a family of mammals or birds is. Indeed, a hive of bees can almost be considered as one individual—a sort of superorganism. This close relatedness encourages the evolution of systems of communication. Furthermore, the system von Frisch proposed is not open-ended. It is strictly limited to information about distant objects of two kinds: sources of nectar and possible new homes. (Swarming bees use their dances to communicate potential hive sites to each other.) The honeybee cannot adapt its communication system to say something about another hive of bees, or the bearing of an object best avoided, or anything else at all.

Adrian Wenner and Patrick Wells of the University of California, Santa Barbara, have spent many years criticizing von Frisch's methods and conclusions. Wenner and Wells's criticisms take several forms. They note that the dances of returning forager bees take place in total darkness—so how are the other bees supposed to observe them? Rather few bees are recruited by these dances, and many new bees find their way to a nectar source without attending to the dance of a bee who has already been there. (Wenner captured every honeybee that tried to return from a food source to the hive and put it in a box where its dance could not be seen by other bees. Nevertheless, plenty of new bees went to the nectar source.) Bees allegedly recruited by following a returned bee's dance are in the air much longer than would be expected if they really knew the bearing they needed to fly on and were making a "beeline" (sorry) for the nectar source.

Wenner and Wells's alternative to von Frisch's dance hypothesis is that bees actually find nectar sources just by picking up the odor that a returning bee carries on its body and then flying off to find that odor. Although Wenner's experiments certainly suggest some role for odor, the most recent data seem to back von Frisch's original observations. Foraging bees really can pick up information from their dancing colleagues. Axel Michelsen and colleagues at the University of Odense in Denmark have built an ingenious, computer-controlled, mechanical dancing bee. With this artificial bee, Michelsen can instruct forager bees to fly off in different directions, without any risk of their being confused by odors or any other factor. Michelsen found that forager bees attended well to the directional information that the mechanical dancer was giving them but tended to ignore its information about distance and always foraged close by the hive. He is optimistic that, given an improved robot bee, he will be able to get the forager bees to fly the distances he instructs them to.

Thus, it does not seem that smell plays a major role in communication among forager bees about good nectar sites, as Wenner and Wells thought. But, in other situations chemicals are very important aspects of honeybee communication. Charles Butler, writing in 1609, knew that the smell of a bee sting would encourage other bees to attack: "When you are stung, . . . yea though a Bee have strucken but your clothes, specially in hot wether, you were best be packing as fast as you can: for the other Bees smelling the rancke savour of the poison cast out with the sting wil come about you as thicke as haile." We now know that honeybees release more than thirty specialized chemicals to influence each other's behavior. These chemicals are known as pheromones and many of them have characteristic smells recognizable to the human nose. The "rancke savour" noted by Butler is usually described today as a banana-oil scent and is a pheromone released when a bee stings, in order to attract other bees to join in the attack.

The queen releases several pheromones that control the activity of the hive. The most important of these is known as queen substance. It is queen substance that prevents workers from laying eggs and keeps them all "cheerefull about their worke," as Butler noted. It inhibits workers from building queen cells, attracts drones to virgin queens on their mating flights, and stimulates normal foraging behavior, the rearing of young, and the building of fresh comb. As Butler correctly observed, the queen is accompanied at all times by six to eight courtiers. What he didn't know is that these courtiers continuously lick the queen and examine her body with their antennae. The queen substance picked up in this way is distributed throughout the colony. If the queen is removed, the workers notice the decline in queen substance within about thirty minutes and become agitated. When queen substance dips below critical levels, the nursery workers start building queen cells and provide royal jelly to rear a new queen. Queen substance is also an "anti-aphrodisiac" that prevents workers from laying eggs themselves. Other pheromones emitted by the queen stimulate brood rearing, comb building, and foraging.

But it is not just the queen who releases pheromones. Workers release Nasonov pheromone (named after its nineteenth-century Russian discoverer) to mark food or water sites for other foragers and at the entrance to the hive to help stragglers find home. Smelling Nasonov pheromone also induces other workers to release their own Nasonov pheromone.

Foraging bees use long-lasting pheromones to mark productive flowers. But they also mark the flowers from which they have just removed all the nectar with a short-lasting pheromone, so that they and their hivemates do not waste time visiting empty flowers. This odor is deposited on a flower during a visit, and it repels other foraging bees for up to an hour. Researchers in England found that this was just the time it took the yellow sweetclover flowers the bees were foraging on to refill with nectar.

Even death is not the end of pheromone communication. The decaying corpse of a bee produces oleic acid, which triggers undertaker bees to remove the corpse from the hive.

Odors are so important to honeybees that they smell in stereo. They have two noses (chemoreceptors), one at the end of each antenna. By having two chemoreceptors in different positions, bees are able to detect the direction that an odor comes from—very important for finding pungent flowers or the source of a pheromone signal. Besides responding to pheromone signals and flower odors, bees can also detect sugar, and they can even tell the difference between different types of sugar, such as fructose, glucose, and sucrose.

Besides smell, bee nerves are sensitive to mechanical disturbances like touch or the force of gravity. Worker bees use texture sensitivity to build comb and distinguish different species of flower—even the different parts of petals. Bees are also sensitive to temperature, carbon dioxide, water vapor, sounds, and magnetism; and they possess neural mechanisms to maintain a sense of the time of day.

Honeybees have five eyes—yes, five: three small eyes known as ocelli on the top of the bee's head and two larger compound eyes at the front. Each of the ocelli has a single, poorly focused lens. Their purpose remains uncertain: it is possible that they react to changes in overall illumination level and adjust the sensitivity of the main bee eyes to maintain optimal responsiveness. Each of the bee's main compound eyes is made up of sixty-nine hundred small sections, known as ommatidia, each with its own lens. It was once believed that the compound eye produced sixty-nine hundred little images, like a kaleidoscope. But more recent research has shown that bees recognize images much as we do, though with only one tenth of our acuity. Honeybees have color vision, with receptors for green, blue, and ultraviolet. The presence of ultraviolet sensitivity means that flowers that to our eyes look plain white may contain significant patterning when viewed by a bee. These patterns may help honeybees identify individual flowers.

This feels like an epiphany. Honeybees, modest, buzzing insects, turn out on closer inspection to have almost mystical powers of perception, navigation, and communication. Yet thus far there is a limitation in everything we have seen honeybees doing: it was just their instincts, their genes, driving them on. Can they go beyond this?

HONEYBEE LEARNING

Though honeybee brains may be small, and much of what honeybees do may be directed, if not dictated, by their genes, nonetheless bees can learn things, and they can even demonstrate simple forms of reasoning. We might almost say that bees engage in abstract thought.

Individual honeybees can learn to associate the odor of a flower with the desirable nectar it contains in a single three-second visit to a flower. They can learn about colors just as quickly. Their ability to find new flowers and return to them implies that bees can recognize and remember shapes and scenes. However, careful experimental studies have shown that a bee's memory for scenes is fairly blurry: the resolution of the image remembered is about ten times less clear than the image formed by their eyes.

Honeybees learn to feed on particular crops at those times of day when the flowers are at their most open. Bees that forage on alfalfa flowers learn to extract the nectar without getting bashed by the spring-operated structure that the flower contains. Inexperienced bees start by sticking their tongue into the open corolla of the alfalfa flower. This releases the flower with a snap that can trap the poor bee's tongue. More experienced bees insert their tongues between the petals at the base of the flower and thereby improve their foraging efficiency and avoid getting whacked.

A combined research team from the Freie Universität in Berlin and the Australian National University in Canberra has shown recently that honeybees do not learn just about physical objects but also about the interrelatedness of objects. Specifically, these researcher teams have demonstrated that bees can master the concepts of "sameness" and "difference." In the initial phase of an experiment, individual honeybees were taught that if they first saw a set of vertical lines on a cardboard disc, then later, when given a choice between similar vertical lines and some horizontal lines, they should fly toward the vertical lines to obtain a sugar-water reward. They also learned that, if they were initially shown horizontal lines, then they should choose horizontal lines later in preference to vertical ones to get the treat.

This pattern of choice—"see vertical first, choose vertical later; see horizontal first, choose horizontal later"—might suggest that the bees had learned that they should always choose the disc containing the same pattern as the one they saw first. However, to show that the bees really have grasped a general concept of "sameness," it is important to demonstrate that they can generalize from the vertical and horizontal lines on which they were trained to some other patterns. Bees that had learned what to do with the two sets of lines suddenly found themselves confronted with blue and yellow discs instead. Now, to get the sugar-water reward, they had to choose the blue disc after first meeting with blue, and the yellow disc after first meeting with yellow. Both German and Australian bees readily and spontaneously chose the color that matched the disc they were first shown. Furthermore, the bees could do the same trick with different odors, and they could also learn to pick whichever item they had *not* been shown originally. In other words, they grasped the concept of "difference" as well as the concept of "sameness."

THE HONEYBEE AND THE HIVE—A REVERSIBLE FIGURE?

So individual honeybees certainly have their abilities. They navigate and communicate; they even demonstrate a simple form of abstract reasoning. The honeybee machine may be small, but it is certainly complex. It has numerous sensory abilities. It can communicate critical information to other machines of the same make and model. Yet a bee on its own is useless: worse off even than a human cast out by his society. Without a hive, a worker would freeze at night or overheat by day. Without hivemates, a celibate worker would have no chance of genetic immortality—no way of passing on her genes. Even if she could lay an egg (which she might, freed of the inhibiting influence of queen substance), there is no way that egg could survive without the protective cells of the hive and nurse workers to take care of it. The lives of the bees in a hive are so tightly integrated that some authors talk of the hive as a superorganism. Perhaps we could understand honeybees better by focusing on the hive rather than the individual bee.

Just as the body of an animal is made up of cells that have some limited ambit of independent action but dedicate themselves to the greater good of the organism, so individual honeybees are nothing without their hives. Workers without their queen, as Butler put it, "never after prosper." As cells are to the ordinary organism, so bees are to the superorganism of the hive. The social intelligence of the hive achieves things no individual bee could do on its own.

All the honeybees in a hive work selflessly for the greater good of the community. There are workers who nurse young, clean house, forage, unload foragers, or even work for the air conditioning detail. (Legions of bees stand near the entrance and fan the air with their wings: bees on one side of the entrance fan fresh air inward, while those on the other side fan the exhaust air out. Worker bees also place droplets of water throughout the hive. As these evaporate, heat is removed, just as in an evaporative air conditioner.)

But just as essential as the positive work of feeding the hive and tending its young is the defensive work of protecting the hive from invasion. Guard bees are specialized workers who wait at the entrance to the hive and keep a lookout for foreign-smelling or -looking bees and attack them. In the ensuing battle, one or other bee may be stung to death. If it survives a couple of raids, the robber bee is less likely to be detected (perhaps because it has picked up some of the smell of the hive) and may sneak in along with the hive's own foragers. Guard duty is important: honey robbing by alien bees may reach such a level of intensity that the defenses of a hive collapse, all the honey and nectar are removed, and the hive ceases to be a going concern.

The first thing I learned about honeybees is that they sting. But the sting of a bee, even of many bees, is not effective against all attackers. In Japan, worker bees attack invading wasps by clinging onto the invader in large numbers to form a ball. The bees then shiver to raise the temperature inside the ball to levels over 110°F. These temperatures are high enough to kill the wasp but do not harm the bees.

In many warmer parts of the world, honeybee hives suffer from attack by a small beetle aptly known as the small hive beetle. This beetle feeds on bee larvae and also on honey and pollen. The bees recognize the aggressor but, because of the beetle's strong exoskeleton, stinging is ineffective. Instead, the bees incapacitate the beetles by imprisoning them in tree resin. A small contingent of foraging worker bees brings back propolis, tree resin, as a building material for the hive. In Africa (though not, apparently, in other parts of the world where bees are under attack from beetles), the bees trap the beetles in corners of the hive and gradually encapsulate them in propolis—a process that can take four days to complete. Even after capture, the bees continue to guard the propolis prisons to make sure no beetles escape. This form of imprisonment is an effective control mechanism against beetle attack.

Though individual honeybees die all the time (only the queen lives more than a few months, and even she is unlikely to survive more than three years), the life of the hive as a community may go on indefinitely. Even if a hive becomes too crowded, or unattractive for some other reason (perhaps because of beetle attack), the citizens of the hive can move off to a better site. A special "whir" dance begins, which incites the majority of the bees in the hive, together with their queen, to swarm. The swarm quickly finds a temporary resting spot, such as a tree limb, from which scout bees, who would normally be out looking for nectar, start searching for suitable sites for a new hive. When they return to the swarm, their communicative dances, instead of indicating the location of productive flowers, indicate sites they consider suitable for a new home. Other workers fly out to check these spots for themselves. If they approve, they add their dance to that of the original scout who found the site, pointing in the same direction. Once enough bees in the swarm reach a sort of consensus (are pointing on the same direction through their dance), they fly off and build a new home. It's a dancing democracy.

An overcrowded hive may divide into two hives in this way—one could say that the superorganism of the hive has "reproduced." It's as if the hive has a life beyond that of the individuals its cells, almost like a body. In the wild, honeybees nest in holes in trees or walls. Once they have settled on a hive site, the bees construct sheets of comb using wax secreted by worker bees and formed on their mandibles and legs into hexagonal cells. Though there are thousands of small cells for workers and drones, only ten to twenty queen cells are built at any one time. The eggs are laid in brood cells in the deepest, warmest part of the hive. Cells around the brood contain pollen (bees harvest pollen for protein). Honey, the honeybees' winter food, is stored in the outermost cells of the hive.

Viewed as a superorganism or just as a rich collection of little organisms, it is an elaborate and complex structure, the hive, with all these insects inside it doing their myriad tasks. How does it work?

On the one hand, we have the individual honeybees, who are capable of far more than one might have imagined but who nonetheless have brains only the size of a sand grain. On the other hand, when a few thousand of these insects are gathered together in a hive, they appear to master feats of social integration that surpass even the most advanced human societies. How is it done?

The first thing to realize is that the queen does not directly regulate the activities of the workers. On this, Charles Butler, writing in the early seventeenth century, was wrong. He saw in the hive a lesson from God on what he took to be the perfect form of government, feudal monarchy. Sure, the queen emits powerful pheromones that influence the behavior of her workers, but she does not send out commands like "Not enough food—send out more foragers," or "Hive getting too hot—put more workers onto air-conditioning duties." Good Queen Bess had considerably more direct influence over her subjects than does the queen bee.

The Division of labor within the hive is influenced by two factors. First, it appears that bees serve their apprenticeships inside the hive before going out into the world. Younger workers tend to stay home; older ones forage outside the hive. But there are also genetic differences between workers who carry out different tasks. This might explain why the queen mates with several drones on her nuptial flight. From one point of view, the hive would be a tighter-knit community if the queen only mated with one male: then all her daughters would be supersisters, share 75 percent of their genes, and be under the most intense evolutionary pressure to help each other. But by mating with several males, the queen picks up for her offspring genetic tendencies toward different behaviors. This variety of behavioral styles may help ensure that her hive gets all the different kinds of workers it needs.

These workers also respond to changing circumstances on a moment-to-moment basis: how is that achieved in the absence of direct control by the queen? It seems that worker bees are motivated by their sisters and by their own experience of the world. Unemployed foragers who observe a particularly enthusiastic dance by a returning scout may be animated to fly out to that nectar site themselves. Scout foragers only dance to their sisters if they have found a food source better than the average of those presently available to the hive. This ensures that foragers are only recruited to worthwhile sites. So that explains how followers choose which of their scout sisters to follow, but it raises a new riddle: how can an individual honeybee know whether the nectar she has returned with is better than the loads that her sisters have recently brought back?

The answer lies with the worker bees who stay home and unload the returning foragers. These bees interact with many different foragers and have a good opportunity to develop a sense of the quality of the nectar "market" at any point in time. When a forager returns with a better-than-average load of nectar, the porter bees unload her very promptly. She is then most likely to perform a dance to let other bees know where she got this good nectar. On the other hand, a forager bee that returns with a poorer-than-average load is likely to have to wait around for some time before a porter deigns to unload her. She is consequently less likely to dance for her fellows. A returning forager finds a waiting time of up to forty-five seconds acceptable—if unloaded within this time, she is still likely to dance; anything longer and she may not. Some returning foragers, however, dance regardless of how long it takes them to be unburdened. The arrogant exhibitionism of these foragers is due to something they know that the bees that unload them cannot: how easy it was to acquire this load of food. A bee that has found a very abundant and nearby source of food may insist on dancing even if she is not promptly unloaded.

While Butler, writing in 1609, saw in honeybees a feudal society, Thomas Seeley of Cornell University, writing in the late twentieth century, observed that foragers "take full advantage of the power of a competitive market process." The dance floor of the hive, Seeley observes, is a "sort of labor clearinghouse where employed foragers advertise their work sites and unemployed foragers listen to these announcements to find suitable jobs."

This then is the key to how the honeybee colony can achieve so much when its members are so limited in their brain power: nobody is trying to understand what is going on. Even the queen, though she has power of a sort through the pheromones she emits, has no overview of the activities of the hive. She is just a citizen herself, albeit a rather special kind of citizen.

Everything that honeybees do is a consequence of three factors. First, the genes that they are born with. These genes ensure that the bees do the right thing at the right time; they can be thought of as a book of rules that tell the individual bees what action to take under what circumstances. Then there is the influence of pheromones. Worker bees stay with the hive, attack intruders, avoid some flowers and attend to others because of special chemical signals that their comrades have released. These pheromones are a straightjacket forcing the bees to perform certain acts. The third layer on top of this genetic and pheromonic determinism is a not inconsiderable ability to respond directly to circumstances: bees fly out to find nectar when they see a particularly compelling dance by another forager; they collect pollen when they feel a protein hunger; foragers who bring a particularly rich load back to the hive get unpacked before their less successful coworkers. And all of this takes place in the context of an intense evolutionary pressure to assist the siblings who carry three-quarters of their own genes. Add these factors together and you have a recipe for a very tightly knit society that acts, as a unit, with amazing intelligence.

ROMANCING THE HONEYBEE MACHINE

Honeybees are small and useful. They are not cute. They also sting. It is hard to imagine a little person hidden inside a honeybee. For these reasons bees have seldom been romanticized. Of course, much that we now know about the complexity of honeybee lives has only become apparent in the last fifty years. The dance language by which bees communicate information about food sources and suitable nesting sites; their complex and highly regimented social lives in the hive; their sensory abilities and capacities for learning and memory—these facts of honeybee life were unknown to Butler in 1609. Indeed, most remain unknown to the wider populace today. This restricts the opportunities for anthropomorphic explanation, as does the honeybee's small size.

The scientists who study honeybees today view the bees' activities as puzzling in the way that a riddle or engineering problem is puzzling, but not as deeply mysterious. Scientific papers on bees show their authors continuously remarking that the discovery they report is another piece in a puzzle much of which still remains cryptic. But there is an implicit confidence that they, or some other scientist, will fill in the next piece, and the next piece, and so gradually the machinery of the bee will be completely deciphered. With its limited, countable, network of nerve cells (even if the count is around a million), its similarly restricted genome, and the approximately thirty pheromones that afford chemical control of the behavior of one bee by another, the machine that is the honeybee is confidently held by all the researchers involved to be comprehensible to the human mind.

As the chemical identity of a pheromone is uncovered and matched to its behavioral effects on the bees that sense it, so the next team of researchers will find the proteins on a membrane somewhere that are changed by the pheromone and speculate about which neural ganglia are likely modified in their activity and how that might in turn cause the behavioral changes that are observed. Though honeybee society may be highly complex and stratified, the existence of other species of bee with less complex societies (such as the bumblebee) and solitary bees (like the Mason bee) makes it possible to conceive of how the very elaborate societies that we see today could have evolved. It is still a riddle how honeybees could have developed the habit of relating the sun bearing on which they must fly to reach a food source into an angle to the vertical at which they dance in the darkness of the hive. But some insight into how such a system might have evolved is offered by the discovery of Asian bees that dance on the horizontal top surfaces of their hives, where they can see the sun and orient to it directly.

Because honeybees are small, not very cute, and because only a few specialists know the most interesting facts about their behavior, bee life is seldom romanticized. Still, some observers would like to interpret honeybees as more than just small, complex machines. Harvard zoologist Donald Griffin, as I mentioned at the outset of this chapter, argues that the communicative abilities of honeybees must be mediated by consciousness. But surely, in our information age, it should be clear that, although conscious beings are capable of communication, communication and information are by no means restricted to conscious entities. Electronic computers and other complex information-processing machines can communicate with each other, yet few observers would want to claim that they are conscious. Honeybees are no closer to being conscious because they can communicate in a highly constrained form than is my computer conscious for being able to "establish protocols" with a distant server so that I can send email messages around the world. Conscious entities communicate, but not all communicating entities are conscious.

The attentive reader will have noticed that I have used a lot of mentalistic language in describing what honeybees do. I asked whether a honeybee would "think" that a journey was farther because a strong wind had made it more tiring. In the same context I said that researchers could uncover how far a honeybee "believed" she had traveled on the basis of her communicative dance. I also said that honeybees search for sites they "consider suitable" for a new hive, and time and again I said that bees "know" many things (time of day, points of the compass, and so forth). I have used the language of "thinking," "believing," and "knowing" not as a backdoor concession that honeybees consciously think, believe, or know the way we do but just because this way of putting things is the most natural language we have for describing such complicated behavior. When speaking of bees, I use these terms the same way I use them when I speak of the temperamental computer on which I work. I know that when I say that this computer doesn't "know" how to spell *Apis mellifera*, what I really mean to say is that a file of word spellings against which all the words in my text are compared when I press F7 does not contain the sequence of letters *Apis mellifera*. I know, too, that this computer doesn't really have moods and good days and bad days. If I'm forced to confront the issue seriously, then I concede that my computer's "bad mood" means only that the extremely lengthy and complicated lists of simple instructions that make up its software occasionally interact in ways that their designers did not anticipate.

The bee, just like the computer, contains a complex machine that processes rules. In the bee's case, just as in that of other animals, humans included, most of these rules are coded in the genes and in neurons. The genes direct the construction of proteins out of which, through complex processes of give and take with environmental influences, the organism is built. Somewhere along the line, neurons are constructed, and it is those neurons that direct the behavior. In the development of neural connections and again in the control of behavior by neurons, environmental interplay shapes the patterns of behavior that will appear. The end result is behavior considerably more complex than anything this dumb manmade computer of mine can do.

These patterns of behavior are so rich that it is easiest to abridge the description of what the bee does and just say it "knows" and "believes" things. But when I do this, I am only really redescribing what the bee does. A forager bee that dances after being unloaded promptly by a porter bee does not really "know" that it has brought home nectar richer than the average cargo that its sisters are bringing in at that point. All that forager has is an algorithm for dancing: "If I am unloaded within forty-five seconds, dance; if not, forget it." The porter bee doesn't "know" that she should unload well-laden foragers first. She just has an algorithm for unloading: "Unload the best-laden foragers first." Even here, where I have written these algorithms in human words that we perceive consciously, there is a danger that I might be seen as supporting the idea that the bees are conscious. I am not. I write these algorithms in English because it's convenient to me. With a little more effort I could have written them in computer code. In the bee's brain they are coded in the connections of the neurons; they are not conscious ideas.

The realism of most observers of honeybee behavior is undoubtedly a good thing. The painstaking progress of research groups around the world, slowly and carefully decoding the next piece of the riddle of bee complexity, is a paradigm case of the sciences of animal behavior at work. But how fantastic the honeybee is. What wonderful machinery evolution builds. One could almost wish for more romance.

FURTHER READING

The World History of Beekeeping and Honey Hunting, by Eva Crane (Routledge, 1999). Everything you ever wanted to know—and many things you would never have guessed—about the history of people and honeybees. An astonishing compendium.

The Wisdom of the Hive: The Social Physiology of Honey Bee Colonies, by Thomas Seeley (Harvard University Press, 1995). An excellent, thorough, and enlightening account of much modern research on life in the honeybee hive.

The Dance Language and Orientation of Bees, by Karl von Frisch (Oxford University Press, 1967). The original—and still the best and most fascinating—account of the dance communication system of the bee by the man who discovered it.

A Book of Bees:—and How to Keep Them, by Sue Hubbell (Houghton Mifflin, 1998). If you think the beekeeping life might be for you, read this beautiful personal memoir of a beekeeper's year.

3

How Noble in Reason

Tenerife is one of the Canary Islands, an island group in the Atlantic about two hundred miles off the west coast of Africa. The Canaries today are as popular with northern Europeans seeking winter sun as is Florida, which is at about the same latitude, with tourists from the northern American States and Canada.

Back in 1913, long before passenger jets, when a young German scientist, Wolfgang Köhler, set out with his family for the Prussian Anthropoid Research Station on Tenerife, the island was considerably more remote than it is today. Their isolation was to get worse. Soon after their arrival the First World War broke out, and the Köhlers were trapped indefinitely by British naval blockade.

Tenerife was surely not a bad place to sit out the war. Köhler found himself with nine chimpanzees of varying ages, some handy enclosures to test them in, and plenty of time on his hands. He proceeded to carry out a series of experiments on reasoning in the chimpanzee that have become classics in the field. After six years of enforced leisure on Tenerife, Köhler returned to Germany, where he became director of the Psychological Laboratory at the University of Berlin—the foremost laboratory of its kind in Europe at that time. With the rise of the Nazis, Köhler left Germany for the United States, where his fame continued to grow. By the time of his death in 1967, he was recognized as one of the most important figures in twentieth-century psychology. I'll come to the experiments that made Köhler famous in a moment. First, I want to consider a study that was carried out *on* Köhler, not by him.

In the 1970s Ronald Ley, from the University of Albany in upstate New York, decided to dedicate his sabbatical leave to tracking down the site of the original Prussian Anthropoid Research Station on Tenerife. Ley was to find a lot more than he had bargained for.

There are several rather odd features of the Prussian Anthropoid Research Station. First, why did the Germans put their ape laboratory on Tenerife? Tenerife was not then a German possession but a Spanish one (as it remains today). Why didn't they put the lab somewhere on their own territory? Second, chimpanzees are not native to Tenerife (how could apes have made their way out into the mid-Atlantic?). The animals Köhler worked with had to be shipped from the Cameroons, several thousand miles to the south along the West African coast. The Cameroons were at that time a German colony. Why didn't the Prussian Academy of Sciences send Köhler there or somewhere else in Africa where apes are native? For that matter, since his research work was done entirely on caged animals, why not just send him to a German zoo? The costs of sending out a researcher with his family and nine chimps to an isolated island were very substantial. An additional surprising fact greets every reader of Köhler's account of his work with the apes: "practically all the observations [in this book] were made in the first six months of 1914." What did Köhler do with the rest of his time?

Tenerife may not be an obvious spot for research on reasoning in chimpanzees, but it is an excellent location for spotting ocean traffic between Europe and North America. The United States didn't enter the First World War until 1917 but well before that had begun to send supplies across the Atlantic that were critical to the allied war effort in Europe. World War I was the first in which submarines played an important role, and German submarines in the Atlantic sank large quantities of allied shipping throughout the war.

Submarines then had nothing like the technology of submarines today. The early German U-boats were only able to submerge for relatively brief periods and could only travel slowly even when on the surface. Furthermore, they needed regular refueling. At first it's hard to imagine how they could have tracked down and destroyed ships on the open ocean. It turns out that the German government, under cover of an astronomical observatory, placed a team of spies on Tenerife. Up on Mount Teide, the highest peak on Tenerife, they watched shipping out at sea and radioed back to Germany. That the "astronomers" were spies Ley was able to prove from contemporary documents in the German archives. Ley deduces from evidence he collected that Köhler was also a spy for the German navy.

That Köhler was acquainted with the "astronomers" is testified to by the recollection of Köhler's children and some photographs that survive from the time. The really interesting question is how much of a spy Köhler was himself. Some evidence suggests that Köhler may have had his own radio transmitter. (Ley is able to demonstrate that he had electricity—quite a rarity for a private home on Tenerife at that time.) He might have used this to communicate with submarines that surfaced just off Tenerife at night. He might even have radioed messages from the submarines back to Germany. The houses where Köhler lived on the island certainly had wonderful views out to sea, and he may have reported sightings of allied ships back to Germany, or directly to submarines.

Patriotism may be the last refuge of a scoundrel (according to Dr. Johnson), but I don't find it at all disreputable that Köhler did what he could, as an expatriate trapped away from his fellow countrymen during a very difficult time, to support his country's war effort. Submarine warfare in those early days was a surprisingly gentlemanly pursuit compared to what it was to become later. Submarines would not fire on unarmed merchant vessels until they had given the crew a chance to abandon ship. They would even tow the lifeboats to a safe landing.

It is a curious thought, though, that some of the most famous experiments in the history of animal psychology may have been carried out to provide a front for a spying operation.

Let's consider for a moment how Ley reached his conclusion that Köhler had been a spy. Ley describes how he went to Tenerife and interviewed people who knew the Köhlers. Ley talked to the West German consul and examined the houses where the Köhlers had lived. He also made trips to West and East Germany, where he examined records from the secret service (the East Germans were surprisingly helpful). Back in the United States he tried to access the Köhler archives (here he had rather little success) and talked to surviving members of Köhler's family. Nowhere did Ley find the "smoking gun"—a definitive document that would prove that Wolfgang Köhler had been a spy. But nonetheless, on the balance of the evidence he was able to obtain, Ley reasoned that that was most likely what Köhler had been doing.

This painstaking collection of facts from which a conclusion is deduced that goes beyond the information given is reasoning par excellence. Could any animal but a human deduce true conclusions from valid premises like this?

In addition to giving a definition of this type—one that Ley, Köhler, and Dr. Johnson would recognize—the Oxford English Dictionary also gives us a more general, weaker sense of what reasoning is: reasoning is the adapting of behavior to solve problems. The OED also says that animals can't reason. It says this because St. Augustine, the Christian theologian of the fourth to fifth century whose writings deeply influenced how the Western world came to understand the difference between people and other species, defined the difference between humans and others as the ability to reason. To St. Augustine, reasoning was an ingredient in the bread on top of the similarity sandwich. Reason was one of the things that separated men from beasts.

I have to disagree. From my own research and my reading of the literature, I place reasoning somewhere in the squidgy filling of the similarity sandwich: it is something that many species of animal can do. Just how similar different animals' reasoning skills are is still an open question. For one thing, the range of different abilities that can be considered under the rubric of "reasoning" is so broad that it would be crazy to expect all animals to have them all to the same degree. But when we look for animals adapting their actions to solve problems and even going beyond the information given to deduce conclusions, à la Sherlock Holmes, we find plenty of species that do these things and do them very successfully.

Animals adapt their behavior to their ends; they choose between alternatives; they solve problems—sometimes fiendishly complicated ones. They reason numerically, conceptually, and deductively. But how do they do these things? I am going to suggest that animals do not make magic leaps of thought but solve problems by stitching together small units, each, on its own, of machinelike simplicity. The result can nonetheless be wonderfully complexity.

But first let's return to Köhler on Tenerife and the research work that made him famous.

BANANA LOGIC

One of the best known of Köhler's problems was the so-called "block-stacking" task. Köhler sets the scene:

> Jan 24, 1914. The six young animals from the research station were locked into a room with smooth walls. The roof—about 2 m high—was out of their reach; and a wooden box (50 × 40 × 30 cm) with one open side was placed roughly in the middle of the room with the open side upwards. The target object [banana] was nailed to the ceiling in one corner of the room about 2½ m along the floor from the box. All the animals tried vainly to attain the target by jumping up from the floor. Sultan soon gave up, paced restlessly around the room, stopped suddenly by the box, grabbed it, tilted it hastily in a straight line towards the goal, climbed on top as it was still about ½ m away horizontally, and jumping up with all his might, ripped down the banana.

So Sultan had reasoned that the box needed to be under the banana before it could be useful as a springboard to reach it. The illustration shows Sultan performing a more advanced version of this test using three boxes that had to be piled on top of each other.

Sultan performing his famous banana trick on Tenerife in 1914. Grande looks on. (Wolfgang Köhler, *The Mentality of Apes*, 1925, Intelligenzprüfungen bei Menschenaffen, © Springer Verlag)

Sultan was also the star of another classic test. This time the chimpanzee was inside a barred enclosure, with some desirable pieces of fruit placed out of reach outside the cage. Sultan was provided with two hollow bamboo sticks of different diameters, each of which was too short to reach the highly desired fruit on its own. At first Sultan tried to reach the food with his hands. Next he tried each of the sticks separately, but neither was long enough for the task. Then he tried nudging one stick with the other but still could not reach the fruit. After about an hour of futile efforts, Sultan gave up. Later Sultan's keeper observed the ape playing with the two sticks in his cage and inserting the thinner of the two inside the thicker one. With this now lengthened double stick, Sultan rushed back to the fruit and raked it all in rapidly.

In all of his studies, Köhler emphasized the animal's sudden change from bafflement to solution. Not for his chimps the blind flailing of trial-and-error learning, the gradual development of a correct solution. No, Köhler's chimps stepped back and pondered the problem for a while before leaping to the correct solution. He felt that the suddenness with which the chimpanzees arrived at the solution to a problem, after a period of apparent reflection, suggested higher mental faculties—in particular, insight. Have you ever seen a chimpanzee stroking its chin, apparently deep in thought? Well, Köhler's experiments are the story that goes with that image.

Well, maybe Sultan was an insightful chimp, not given to fumbling, trial-and-error learning. But then again, maybe not. Köhler's studies, though they made his name, don't get high marks from modern commentators for their experimental methods. We have to contend with the fact that Köhler allowed up to six chimps to work on any one problem at one time (a group of apes is called a "shrewdness"). One chimp might make a few unsuccessful attempts at solving the problem; then a second chimp would go straight for the correct solution without making an error. Had the second chimp had a flash of deductive brilliance without the need for trial and error, or had he just watched the first chimp and learned from her mistakes?

Köhler also tells us so little about his chimpanzees' prior experiences with the materials he provided for each problem's solution. He does mention that the chimpanzees at the Prussian Anthropoid Research Station were given many opportunities to play with packing boxes, hollow bamboo sticks, string, and many other things. We know very little, however, of what the different chimps did with these objects. Quite possibly, Sultan may have previously stood on packing boxes to reach desired items or inserted one stick inside another for some purpose. Köhler may not have seen any trial-and-error learning, but it is hard to be sure that the trials and errors had not taken place earlier, while the apes were playing and nobody was watching. Several groups of later researchers have attempted to unravel what role past experience might play in these apparently insightful chimpanzee leaps of logic. The answer seems to be that appropriate opportunities to play with boxes and sticks are important preludes to the kind of performance Köhler observed in Sultan and the others.

The difficulty of interpreting the reasoning abilities of the Tenerife apes has led to a very different style of experimentation on animal reasoning in more recent times. No longer is a shrewdness of apes offered some riddle in the midst of their familiar play area with a selection of toys lying around that might be used as tools. These days animals are taken out one at a time and tested under very carefully controlled conditions.

In any case, if we're looking for reasoning in animals there is no need to start with our closest relatives, the great apes. As we saw in chapter 2, insects are good for surprises—let's consider reasoning in wasps.

THE STING

Mud dauber wasps are the ones that build the mud "organ pipes" commonly found on the sides of buildings in North America. I remember when I first saw these strange tubes—about a half-inch across and six inches long—on the inner walls of the garage to a house I rented in North Carolina in the early 1990s. It never crossed my mind they had been built by insects; to me they looked like some human construction out of cement. I was rather shocked to learn that mother wasps had built them to house their eggs.

Split open one of these mud cylinders and you will find several immobile, yellowish-green crab spiders inside and, visible with a hand lens, a single egg that the mother wasp has deposited in one of them. The crab spiders are not dead. They have been stung into paralysis by the mother wasp before she stuffed them into the cylinder. By leaving them alive but paralyzed she ensures they will not decay before her egg hatches. The hatching wasp larva will then find a nice, fresh supply of spiders ready to be consumed.

Different species of wasp set up their eggs for life in different ways. Paper wasps build nests of a papery substance and provision their young with incapacitated caterpillars. Digger wasps dig holes for their eggs in the soil and choose grasshoppers, crickets, or beetles (each species of digger wasp has its own preference) to load the nest. One species of digger wasp, *Larra bicolor*, has taken the process a step further. It cuts out the time-consuming hole-digging component of the task (it can take a digger wasp over an hour and a half to dig a suitable hole for one egg) and lays its egg straight into the body of a mole cricket. The cricket is temporarily paralyzed but quickly recovers and goes back about its business. After a week or so, the tiny wasp larva emerges and feeds on the cricket from the inside out. Within a few weeks the grub is full grown, and the mole cricket has been destroyed.

If these gruesome habits sound like something from the more horrific forms of science fiction, bear in mind that such things are probably going on around your house right now. In the eaves, the garage, or the yard, some species of wasp will have laid its eggs and supplied them with a living store of baby food.

The macabre egg-laying habits of solitary wasps were first described by a French naturalist, J. Henri Fabre, in the early years of the twentieth century. To Fabre, living in an age before refrigeration, the biggest puzzle was how the wasps succeeded in keeping the food for their eggs fresh in the summer. Fabre extracted beetles that digger wasps had buried and compared their rate of decay with beetles of the same species that he had captured and killed himself. In his lively work *The Hunting Wasps*, Fabre records his observations: "in a heat which, in a few hours, would have dried and pulverized insects that had died an ordinary death, or in damp weather, which would just as quickly have made them decay and go moldy, I have kept the same specimens [beetles that digger wasps had buried], both in glass tubes and paper bags, for more than a month, without precautions of any kind; and, incredible though it may sound, after this enormous lapse of time the viscera had lost none of their freshness." Through careful dissection (or rather, as it turned out, vivisection) of the stunned beetles, Fabre came to recognize that the poor beasts were not dead, just paralyzed. He carefully watched wasps sting these beetles and made his own experiments injecting poison into the same part of the beetle's body with an old-fashioned pen nib. Sure enough, introduction of poison into the body at just that point destroys enough of the beetle's nerve centers to render it incapable of movement, without killing it outright. Fabre went so far as to carry out simple experiments to show that paralyzed beetles survive longer than do fully active beetles imprisoned without food.

One chapter of Fabre's book is titled "The Wisdom of Instinct." In it Fabre outlines the beautiful ingeniousness and elegant adaptedness of the mother wasp's instinct for provisioning her young. The otherwise vegetarian adult wasp turns to insect hunting solely to secure suitable food for her egg. In the case of the digger wasp, she finds soil of just the right friability and digs a pit of exactly the right depth and width (some species use preexisting cavities, in which case the wasp first checks the hole and digs it out to fit). Then she finds the beetles (or whatever her favored prey may be), stings one in just the right spot, the only part of the beetle's body that will ensure paralysis, flies back to the burrow clasping her prey, checks for obstructions before dragging the beetle by its tentacles into the hole, collects just the right number of prey, lays her egg in one of them, and, finally, collects tree resin to seal the burrow.

The next chapter of Fabre's work is called "The Ignorance of Instinct." "The wasp has shown us how infallibly and with what transcendental art she acts when guided by the unconscious inspiration of her instinct; she is now going to show us how poor she is in resource, how limited in intelligence, how illogical even, in circumstances outside her regular routing." For here's the rub: even quite small departures from the normal run of life leave the wasp baffled and unable to cope. Fabre performed a number of small experiments showing up the wasp's inflexibility. He cut the antennae from a locust that a wasp of one species was bringing home and found the wasp quite unwilling to pick up her captive by any other part of its anatomy. He disturbed a wasp of a different species in the process of blocking off her provisioned burrow and removed the caterpillar that this wasp had chosen as the nursery meal for her young. Notwithstanding that the wasp inspected the burrow after having the caterpillar stolen from her, she nonetheless recommenced the work of blocking off the top of the burrow without replacing the missing caterpillar with a new one.

One species of wasp that prefers locusts always puts down its prey to check the burrow for obstructions before grabbing the locust by its antennae for a final tug into the hole. Fabre found that if he dislodged the locust from the wasp just before she entered the burrow, then the wasp would search her nest anew before proceeding. And she would repeat this inspection for as long as Fabre could find the patience to test her: with each new disturbance, the mother wasp repeated the inspection of her burrow.

One species of wasp that normally likes to stock its burrow with four crickets Fabre found would give up after three or even two if robbed of its prey at the final stage of stuffing the cricket into the hole. It appears that this wasp is driven to hunt and catch four crickets, not to actually ensure that four are buried with her egg.

The moral that Fabre drew from these tales nearly one hundred years ago is the same one I would point to today: "Instinct knows everything in the undeviating paths marked out for it; it knows nothing outside those paths." The process of evolution, naturally selecting those wasps with habits that lead to the survival of more offspring in preference to those whose actions are less likely to ensure their representation in future generations, is not a trivial matter. This form of learning or intelligence, if we want to call it that, is an astonishing product of unimaginable lengths of time and an unforgiving god. But the intelligence of instinct is a brittle one, just smart enough for the circumstances that arise regularly. It has no slack—no spare processing capacity, a computer scientist might say—to pick up unexpected conditions. Consequently, we probably wouldn't want to honor the inflexible ingenuity of wasps with the label "reasoning."

Mother wasps are certainly very inflexible about the kind of prey they will collect as food for their young and what they do with it once they've caught it. Although different species of wasp choose different prey (indeed, entomologists use the choice of food as a sure-fire indication of the species of wasp they are dealing with), wasps of any one species are very loyal to their favorite prey species. But how do mother wasps recognize their prey? It turns out that wasps have an excellent sense of smell, and it is this that enables them to identify the right prey species. Furthermore, though much of the mother wasp's behavior may be rigid and inflexible, her ability to learn about the value of odors is remarkably adaptable.

One small species of wasp that feeds caterpillars to its young finds the caterpillars it needs by tracking odors. James Tumlinson of the United States Department of Agriculture in Gainesville, Florida, together with his collaborator, Joe Lewis of the USDA in Tifton, Georgia, and their respective research teams, uncovered a very strange story of plant-to-insect communication. It turns out that the caterpillars these wasps feed on are odorless. Since the wasps would quickly sniff out a smelly caterpillar, it is not entirely surprising that the caterpillars have evolved to give off no smell. But whenever a caterpillar bites into a leaf, the injured plant gives off a characteristic odor that the wasps track to find the caterpillar. The plant refrains from emitting this odor if it is damaged in any other way. Only the presence of caterpillar spittle in a wound triggers the plant to release the odor that attracts the wasps. Furthermore, the plant does not exude the odor passively or in just its wounded leaf but all over: remove that leaf and the wasp still finds the plant. A highly adaptive case of plant-to-insect communication: the plant is sending a signal that we can decode as, "The caterpillar's over here."

It turns out that responding to distress signals from plants is not the limit of wasps' ability to act flexibly and adaptively. Lewis and his group decided to take some of these wasps into the laboratory and see just how much they could learn about odors. He provided the wasps some attractive sugar to eat and while they were eating exposed them to an odor: Would the wasp later choose that odor over others in a preference test? Lewis and his coworkers found that the wasps were astonishingly flexible in their odor learning. These insects were quick to learn that the smell of chocolate meant sugar water was available—a connection that would surely never arise in nature. Emboldened by these results, Lewis' group decided to try the wasps on the smell of TNT. Now they have wasps that can detect land mines.

Nothing in the wasp's ability to detect the odor of TNT matches the complexity or ingenuity of her behavior in stocking a burrow with stunned prey for the child she will never meet, but, in the search for reasoning among animals, it is nonetheless a great step forward. For here, instead of the genetic wisdom of the ancestors carried forward through the generations and played out inflexibly in each individual, we see the wasp, in some small way, figuring something out for herself. She can learn that here and now, in this little world Joe Lewis and his colleagues have built for her, TNT smells can imply nutritious sugar water. One small step for a wasp; one giant leap for evolution.

Before we leave insects, I want to consider the question of consciousness in wasps. As I mentioned in chapter 2, Harvard zoologist Donald Griffin argues in *Animal Minds: Beyond Cognition to Consciousness* that the communicative system of honeybees is possible evidence for consciousness. Griffin also argues that commentators have overemphasized the inflexibility of mother wasp behavior and that we should not too hastily deny the possibility of consciousness in wasps. But if communication is a true sign of consciousness, and inflexibility of behavior does not disqualify an individual from being considered conscious, should we not consider as equally likely the possibility that the plants that send signals to wasps might be conscious? They communicate—which Griffin views as positive evidence for consciousness—and, on the example of mother wasps, the inflexibility of the rest of the plants' behavior should not disqualify them from being considered conscious. To me this is a *reductio ad arbsurdum* of Griffin's position. Of course we can't consider plants conscious. (If you don't agree that plants can't be conscious, then your concept of what it means to be conscious is just so different from mine that there is little point our continuing to discuss the matter.) We should not be misled into thinking that every example of complex behavior is proof of consciousness; complex behavior can arise from simple mechanical processes. This moral will be repeated several times before this chapter is over.

The ability of animals to connect cues and consequences in the way that the wasp can learn to connect TNT smells to sugar was first recognized by Ivan Pavlov, a Nobel Prize–winning Russian physiologist turned animal psychologist of the late nineteenth and early twentieth centuries. In his classic experiments, dogs learned of signals (such as tones and lights) that predicted that food was going to be placed in their mouths. The subsequent century of research has demonstrated that a very wide range of animal species can learn to associate causes and effects in the world around them, as well as notice the consequences of their own actions. This form of learning (known in the trade as associative learning, because signals and their consequences become associated) is an often underestimated tool of behavioral adaptation. It is extremely widespread: not a vertebrate has been tested that could not link some signals to the events they portend, nor recognize the consequences of its own actions. Many invertebrates and even insects, such as snails, wasps, and bees, have also been shown to be fully capable of associative learning. Evolution has primed animals to be better able to associate cues and consequences that actually mean something in that species' world than completely arbitrary pairings of cues and consequences. The process is more powerful for having this selective character than if it were completely open-ended. Associative learning can also run a surprisingly long way: quite complex outcomes can be determined by such a simple mechanism, as we shall see later.

Associative learning is so widespread that, if you have any pets, you can easily observe it at home. If you have a dog that drools you probably already know that he has learned to associate certain of your activities with getting fed. Dogs are so quick in this regard that most dogs have learned what is meant by "walkies," and some owners have to spell out certain important words if they do not want Fido going crazy every time "walk," "park," or "leash" crop up in conversation. Cats too will readily learn of any signals that predict feeding time. We used to have a cat who reckoned that any use of the can opener meant she was in for something tasty. My father-in-law has taught his goldfish that he is about to feed them by turning on a hose of fresh water in their pond just before bringing out the fish flakes. They show their recognition that food is imminent by swimming up to the pond's surface—something they only used to do after the flakes had already been scattered on the water.

Associative learning is simple and mechanical—simple enough that in some species its mechanisms in the brain are fairly well understood—but its adaptive flexibility also makes it very powerful. Powerful and adaptive enough that it deserves to be considered a form of reasoning. "Powerful and adaptive" yet "simple and mechanical"—this is an apparently contradictory pairing that we shall meet again in our exploration of animal reasoning.

But associative learning can only be considered as reasoning in the broader and weaker of the two senses of that word. Associative learning is reasoning in the sense that it is an ability to adapt behavior to solve problems in the world. It cannot be considered a form of reasoning in the narrower, more Sherlock-Holmesian sense of deducing true conclusions from incomplete information. This is what Wolfgang Köhler had been looking for on Tenerife (that, and allied shipping). I want to consider next some evidence for reasoning in this narrower but more exciting sense.

MODERN LOGIC FOR PRIMATES

The tests of reasoning that impress me most involve fairly simple pieces of equipment without too many moving parts. I like the tests that have no hidden tricks but keep everything that is needed for successful solution out in the open. Before looking at some stunning successes in the quest for animal reasoning, let's first consider two surprising failures.

One telling study tests the ability of capuchin monkeys to figure out how to get treats out of tubes. This is a problem that motivates monkeys as naturally as it would children. Capuchin monkeys are the lively little guys from South America, with curly tails and a cap of dark hair, familiar from their role on the shoulders of organ grinders. They get their name from the Capuchin order of monks: their tuft of dark hair is reminiscent of the monk's cap (the word in Italian is *cappuccino*).

Elisabetta Visalberghi and Luca Limongelli from the National Research Council in Rome tested the reasoning skills of capuchin monkeys on a simple task they called the "trap tube" problem. Their trap tube is a simple straight glass tube about a meter long (three feet). Think of the kind of tubes you can get at the post office to send posters in, but made of glass. Now add a hole and a trap halfway along the tube. One by one the monkeys were shown the tube placed horizontally across a frame with a candy inside it. The monkey was provided with a stick of just the right thickness to push down the tube and dislodge the candy. The challenge was to push the stick down the glass tube from whichever end would ensure that the candy came out the far end without getting caught in the trap in the middle. If the candy came out the far end, the monkey could eat it, but if it fell in the trap in the middle, it was lost. This looks pretty easy. The tube, remember, is transparent, and the trap in the middle is very obvious. a no-brainer, right?

But Visalberghi and Limongelli found that, even with extensive testing, only one of their four monkeys learned to consistently push the candy from the direction that ensured it came out the end and did not fall down the trap. Even this one clever monkey did not seem to have a very deep understanding of what was going on. If the tube was rotated so that the hole now pointed upward and consequently nothing could fall out, the monkey continued pushing the candy with the stick from the far end of the tube, as if the trap were still an obstacle to be avoided. No great insights here; no evidence of deductive reasoning.

One of Elisabetta Visalberghi and Luca Limongelli's capuchin monkeys gingerly inserting a pole into a tube in an attempt to push a candy out the other end without letting it fall down the trap in the middle. Candies that got caught in the trap were not given to the monkey.

When the same task was presented to a bunch of chimpanzees (again, tested one at a time), performance was a little better. This time two of the five chimpanzees were successful and learned to consistently push the candy so it wouldn't fall through the hole. To test how much these two chimps really understood the task, they were offered a new version of the trap tube. This time the position of the hole could be moved closer to one or other end. Confronted with this modified version of the tube, only one of the chimps was consistently able to insert the stick into whichever end of the tube would ensure the candy was not lost down the hole. This one highly successful chimp was Sally Boysen's Sheba, who, as we shall see later in this chapter when we discuss numerical reasoning, is clearly an exceptionally gifted ape. The second chimpanzee was also successful much of the time, but only by gingerly inserting the stick into the wrong end of the tube, and then pulling it out and trying again from the other end when he could see that the candy was about to fall through the hole.

Another elegant and straightforward test of reasoning in monkeys involves tubes. This time forget rigid transparent tubes and think of flexible opaque tubes—the plastic kind you might use to fix a gutter to a downspout. Now construct a frame that offers three positions at the top for one end of a tube and three positions at the bottom for the other end. Attach one end of the tube to one of these three inlets at the top and then connect the bottom end to one of the three outlets at the bottom. You can connect the tube however you like—so long as it is bent. In no case may the tube be connected to the outlet directly beneath an inlet. The illustration shows the apparatus in action.

The apparatus used by Bruce Hood and colleagues to investigate cotton-top tamarins' abilities to figure out how a candy would fall through a bent tube. The candy was dropped into the frame at the top of the tube, and it fell out (of course) at the bottom end of the tube. With the box set up in the arrangement shown here, not a single tamarin guessed correctly how the candy would fall.

Cotton-top tamarins are another species of lively New World monkey named for their characteristic (in this case white) head coloring. Bruce Hood and colleagues at Harvard University tested these monkeys on a very simple task using the tube box sketched above. Each tamarin was given the opportunity to observe objects being dropped down a tube in this box. All the monkey had to do to obtain food was point to the spot where the object would fall out. Just to reiterate the rules: at all times the connectedness of the tube was in plain view; nothing was hidden from the monkey. But despite the task's apparent simplicity, the tamarins seldom pointed to the right spot. Even after extensive training, each tamarin remained convinced that the object had to fall straight down. They never got it that the object would be forced into a detour by the tube. No evidence of reasoning here; no adaptability of behavior.

I mention these two failures of monkey reasoning first because I'm now going to mention a number of successful demonstrations of reasoning in animals, and I didn't want you to hear about them without first knowing that there are substantial limitations to how nonhuman species reason about the world. As we shall also see in chapter 7, to nonhumans, seeing is not necessarily believing.

WHO IS SMARTER?

If we want a pure test of deductive reasoning, maybe we should head back to the ancients who first identified the different forms of logic. That's what Cyril Burt, a British psychologist of the first half of the twentieth century, did when he set out to create the first British IQ tests. He remembered his own schooling on tests of logic from Aristotle and decided to use a few of them and see how well they worked as measures of intelligence. One problem that he found was particularly efficient at separating the slum boys (as he called them) of Liverpool from the finer minds at an Oxford preparatory school went like this: "If Mary is taller than Susan, and Susan is taller than Jane, who is the tallest, Mary, Susan or Jane?" Now that's real reasoning. To solve this problem, the children had to deduce new knowledge by going beyond the information provided. They were told two facts (the relationships between Mary and Susan and Susan and Jane) and had to deduce a third (the relationship between Mary and Jane). This particular form of deduction is known as a transitive inference, and it is an attractively clean test of reasoning.

Few people know that one of the first jobs the great founder of developmental psychology, Jean Piaget, had as a young man was translating Burt's English IQ test into French. (Piaget was a francophone Swiss.) Piaget became very interested in the transitive inference problem for what it could tell him about the development of reasoning in the child. He determined that only children who had reached a fairly advanced level of logical reasoning that he called "concrete operational" thought could solve problems of this level of complexity. Concrete operational thought usually kicks in around age seven.

Some child psychologists wondered if children younger than seven years might be confused, not by the underlying logical structure of the transitive inference task, but by the way the problem was being put to them. If you ask a four-year-old child, "Who is the tallest, Mary, Susan, or Jane?" she will have difficulty freeing herself from what she knows about real Marys, Susans, and Janes of her acquaintance. Perhaps her older sister is called Jane, whereas her best friend, Mary, is a dwarf. In real life Jane is taller than Mary, not the other way round as our transitive inference riddle insists. Smaller children may have difficulty breaking free from the knowledge that they have of the world around them and accepting the arbitrariness of the Marys, Susans, and Janes in psychologists' experiments.

In the early 1970s Peter Bryant and Tom Trabasso, both at that time at the University of Oxford, decided to try and develop a form of the transitive inference task that small children might have a better chance of understanding. They modified the task so that, instead of being told about hypothetical girls, the children would be presented with some concrete objects to reason about. They were shown pairs of small sticks of different lengths and colors—rather like a set of colored pencils that had been sharpened down to different lengths. In their experiment, Bryant and Trabasso kept the lengths of the sticks hidden from the children beneath a cover. (Imagine these colored pencils placed beneath a tray so that exactly one inch of each pencil sticks out from underneath.) Though the lengths were obscured, the children could still clearly see the different colors of the sticks.

So Bryant and Trabasso showed each child the tips of pairs of colored sticks protruding from behind the cover. A child was shown a blue stick with a green stick and told that the blue stick was longer than the green one. This would be repeated until the child could repeat back, when asked, that, yes, the blue stick was longer than the green one—without ever having been shown the true lengths of these sticks. The child was also trained to report that the green stick was shorter than the blue one. Once that was mastered, the child would be shown the green stick with a yellow one and this time told that the green stick was longer than the yellow one. Remember, in each comparison only one inch of each stick was visible. After enough repetitions of question and answer for the child to comprehend that the green stick was longer than the yellow one, the procedure would move on to yellow and violet sticks and finally to violet and red sticks. In the end the child had been shown five sticks and told that blue was longer than green, green longer than yellow, yellow longer than violet, and violet longer than red. At no stage was the child allowed to see the true lengths of the sticks.

This rather elaborate-sounding procedure is all a way of conveying to the children the relative sizes of a series of objects—just as Burt's schoolchildren had been told that "Mary is taller than Susan" and "Susan is taller than Jane." Bryant and Trabasso's procedure has the advantage of using concrete objects, not imaginary ones, and of not introducing hypothetical language ("if") along the way. At each stage of the procedure, we are dealing with actual sticks set in front of the child, only their lengths being hidden. The final critical test in Bryant and Trabasso's experiment was to show each child the green and violet sticks (with their lengths hidden as before) and to ask which was longer. This time the experimenters praised whichever answer they got; they did not attempt to convince the child that the green one was longer. This presentation of green and violet sticks is equivalent to asking, "Who is the tallest, Mary, Jane, or Susan?" With the problem presented in this more concrete manner, Bryant and Trabasso found that their four-year-old subjects were just as successful as Piaget's seven-year-olds had been when asked about the heights of hypothetical Marys, Janes, and Susans.

Why did the researchers use five objects instead of three, as in the original version of the transitive inference problem? Imagine that they had only used three sticks; "blue is longer than green, and green is longer than yellow." The problem here is that, if we now asked, "Which is longer, blue or yellow?" the child might answer "Blue" simply because, of the two colors on offer it is only blue that she has ever heard referred to as "longer than" other sticks. A set of five sticks gets around this problem because we don't ask the child about the first and fifth sticks in the series but about the second and fourth sticks. When the child is asked, "Which is longer, green or violet?" she is being asked about two sticks, both of which have been "longer than" and "shorter than" other sticks an equal number of times. When four-year-old children answered that they thought the green stick was longer than the violet one, they were forming a transitive inference and reasoning deductively in much the same way as their older brethren did when asked about Mary, Susan, and Jane.

The capital of Scotland is a strange place for a bunch of South American squirrel monkeys to end up. When I arrived in Edinburgh as a graduate student in 1984, Brendan McGonigle's squirrel monkeys were living in the best-insulated accommodation in that drafty city. They had recently completed their first experiments in transitive inference. McG (as he likes to be known) together with his partner in crime, Margaret Chalmers, had recognized that Bryant and Trabasso's adaptation of Burt's transitive inference problem to sticks could be further modified to test whether such an apparently complex form of reasoning might be found in monkeys. McG is an enthusiastic smoker of pipe tobacco, which used to come in small cans. He and Chalmers painted these cans in strong primary colors and presented them (with their lids firmly shut) in pairs to each squirrel monkey. Given a choice between a blue can and a green one the monkey had to learn that, if it dislodged the blue can, it would find a peanut underneath, while under the green can was nothing. But on trials where a green can was presented with a yellow one, it was the green can that hid the peanut. Then again, on trials with a yellow can and a violet one, it was the yellow can that hid the peanut. And, finally, on trials with violet and red cans, the violet can hid the peanut. This is just like the "Blue is longer than green, but green is longer than yellow" relationship that Bryant and Trabasso's children had puzzled over, only this time the relationship is "Blue is better for peanuts than green, but green is better than yellow."

In this way—the monkeys rewarded with peanuts, the lab staff with fish and chips and pints of Belhaven 80 Shilling—the monkeys gradually mastered all of these four distinctions to a high level. The pièce de resistance came on the day when a monkey was first presented with green and violet tobacco cans together. Would the monkey recognize that, because green was better for peanuts than yellow, and yellow was better than violet, then green should be preferred over violet? Great was the jubilation when the monkeys first made their choices and consistently selected green.

Keep in mind that, during the original training, peanuts were (in toto) found equally often under green and violet containers, so from the point of view of how often different choices were rewarded, green and violet should be preferred equally. But instead the monkeys chose green, reasoning, in effect, "Green is better than yellow, and yellow is better than violet; therefore green is better than violet."

McGonigle and Chalmers's success in demonstrating transitive inference formation in squirrel monkeys came as a shock to the small world of people interested in animal reasoning. Suddenly animals were solving a deductive reasoning problem that only a few years previously had been thought the preserve of older children and human adults. How far might the ability to reason logically stretch?

A few years later I made another new start, this time at the University of the Ruhr, in Bochum, Germany. I was met at the train station by Lorenzo von Fersen—whose blue eyes and blond hair are classically German but whose whole manner and demeanor give away his South American upbringing. Lorenzo spoke as little English as I German and so we were assigned to improve our skills in the other's language. Though we fast became friends, our communicative difficulties at first obscured for me that Lorenzo was training pigeons on a modification of McG's transitive inference task for squirrel monkeys. The pigeons couldn't be made to choose between tobacco cans hiding peanuts, but the workshop at the University of the Ruhr had come up with a very clever apparatus that the pigeons could use to choose between visual patterns. These patterns were basically simple doodles that one of the technicians had drawn on a notepad while talking on the phone. The workshop built an ingenious little horizontal work surface, made to be a comfortable size for pigeon use, with two identical sections to it. The doodles could then be projected from underneath to appear in front of the pigeon on either the left or right section of the work surface. If the pigeon liked the look of a doodle it could peck at it and the bird's pecks would be sent to a computer. The computer would then decide whether the pigeon had pecked on the correct pattern and, if so, would arrange the delivery of food grains through a small chute onto that section of the work surface. It was all rather complicated, but the reputation of German technical workmanship is well earned, and the apparatus was highly reliable.

The doodles used as stimuli weren't supposed to resemble anything, but von Fersen came up with names for them just to keep things straight in his own head. I don't remember these names exactly now but they were things like "little star," "keys," and so on. The pigeon might be confronted with "pig head" alongside "little star," and it would have to learn that pecking on "pig head" could get you food grains under that condition. But then it would be shown "little star" alongside "keys" and would learn that this time pecking "little star" could get you food grains. And so, just like McGonigle and Chalmers's squirrel monkeys and the children Bryant and Trabasso worked with, von Fersen's pigeons gradually learned a series of preferences among five arbitrary stimuli arranged in four pairs.

In Bochum too the excitement was great as the pigeons mastered all four pairs of visual patterns and were given the critical choice between the second and fourth patterns in the series of five. Would even pigeons solve the transitive inference problem? Indeed they did. The pigeons chose "little star" in preference to "deformed question mark" (or whatever they were) spontaneously and at a high level of reliability. They could solve the problem just as well as monkeys and young children. This time good German beer flowed freely in celebration.

To find such apparently complex reasoning in monkeys, birds, and rats (as others subsequently did) raises at least two questions. Why should such animals be able to deduce transitive inferences? And how do they do so?

REASONS FOR REASONING

It isn't as hard to see how animals could benefit from being able to form transitive inferences as I at first thought. Though we think of deductive transitive inference as abstract and difficult, it is actually the bedrock of a fundamental principle: the principle of ordering or ranking. Any animal that needs to form rankings of things would benefit from the kind of reasoning embodied in the transitive inference task. A social animal, for example, that lives in a hierarchical society would live a quieter and more fulfilling life if it could comprehend the relationships between all members of its group without needing to test itself against every last one of them. Dorothy Cheney and Robert Seyfarth from the University of Pennsylvania conclude from their studies of vervet monkeys in Kenya (of whom more in chapter 5), that these animals infer dominance hierarchies among the members of their group by observing interactions between other members of the group and using much the same mechanism of transitive inference as McGonigle and Chalmers's squirrel monkeys did in the laboratory in Edinburgh. "If Monkey 2 can beat Monkey 4 like that, and I know from bitter experience that Monkey 4 can beat me, then I'd better give Monkey 2 an extra-wide berth" is the kind of transitive-inferential reasoning that goes on commonly among the vervet monkeys of Kenya.

The same advantages would be true for rankings of food items. If a pigeon prefers maize to wheat and wheat to millet, then given a choice between maize and millet it should spontaneously prefer maize—no need to play around with the millet first. That's transitive inference, plain and simple. And put like that, it's not so surprising to find that pigeons can do it. The same argument applies to the ranking of nesting sites, or any other orderable objects: reasoning of the type shown in the transitive task would be helpful in all these contexts.

So without having to try very hard we can come up with many reasons why a wide variety of species might find transitive inference useful if they could do it. But how do they do it? There's the rub.

One possibility that von Fersen and I explored assumes that each object shown to an animal in a transitive inference experiment has some value to it. Each time the animal picks that object and receives a reward, the object becomes a little more valuable. Each time the animal chooses and finds no reward, that object becomes a little less valuable. If the objects are very similar in value, then the animal might choose at random, but if the objects differ clearly in how valuable they are, then the animal will always choose the more valuable one.

This principle might sound too simple to account for such a complex deduction as the transitive inference problem demands, but it is in fact quite adequate for the task under many conditions. Imagine you are a squirrel monkey in McG and Chalmers's transitive inference experiment. Perhaps not an easy thing to do. On the table in front of you put out five piles of fifty dollars each (Monopoly money will do). These represent the value to you of each of the tobacco cans. In front of you are two cans on a tray: one blue and one green. You know that hidden under one of these cans is a peanut. You like peanuts. What should you do? Well, at the moment both cans have fifty dollars on them, so just pick at random. You choose the green—wrong! No peanut. Deduct two dollars from the money pile for green. Keep going with blue and green, choosing at random. Subtract two dollars from the green pile for every wrong choice, and add two dollars to the blue pile for every correct choice until there is a clear difference in value between blue and green—let's say a clear difference is a difference of ten dollars or more. Once the difference reaches ten dollars, you always thereafter choose the higher valued item: the blue can.

Once you have completed ten blue/green trials, start your green/yellow training. Now you should find that, at first, you have more money on yellow than on green, and so you start by incorrectly choosing yellow. Yellow loses money until you choose at random for a while, and then you add value to green and consistently choose the green can. Give yourself ten green/yellow trials before you start on yellow/violet. At the start of yellow/violet training, you will find that yellow value has been depressed even more than green was at the start of green/yellow training, and therefore you spend even longer making incorrect violet selections. But persevere for ten trials and the situation will straighten itself out. Finally, give yourself ten violet/red trials. If you have the same experience as I did, you should find at the end of fifty trials of training that your money piles end up looking something like this (totals may differ slightly because we won't have chosen the same cans when choosing at random):

Blue	$70
Green	$62
Yellow	$54
Violet	$54
Red	$42

So now you are ready for the critical test of deductive reasoning: the comparison of the second and fourth items in the set, the green and violet cans. Would you be able to choose rationally using this scheme? Yes, you would. Green has twelve dollars more on it than violet, so you would choose green. In fact, a scheme like this (mathematically more sophisticated but using these principles) can emulate very well the performance of pigeons, rats, monkeys, and even humans on transitive inference problems.

Here's the score so far. Simple reasoning about how things will fall in tubes—not even apes get it. Complex deductive reasoning in the transitive inference task—apes, monkeys, pigeons, and rats all get it. But how is it done? With a simple mechanism. Complex behavior achieved by simple mechanisms—a recurring theme.

So what other deductions might a pigeon or monkey be capable of? Herbert Terrace at Columbia University in New York, with a task that looks very similar to the transitive inference problem, has found severe limits to a pigeon's ability to deduce.

In the transitive inference task, pigeons (or monkeys, apes, or people) have to *infer* a series. They are only shown two items at a time, but they deduce relationships among the whole series of items—relationships they have never been shown. Terrace wondered how pigeons might do if they were first shown a whole series of colors and then asked to reason about just a few colors selected from that series.

Terrace and his coworkers showed their pigeons five plastic keys to peck on, each key illuminated with a different color: red, green, blue, yellow, or violet. On each run-through, the pigeon had to peck those five keys in a fixed order of colors—let's say, first blue, then green, then yellow, then red, and finally violet—in order to get a reward of grain. Each time the colors were jumbled anew across the keys, so the pigeon could not learn just to peck the keys in a specific order—say, from left to right. This sounds similar to the transitive inference problem, but the difference here is that all five colors are present at the same time, and the lights only go off once the pigeon has pecked all five in the right order (or as soon as it makes a mistake, such as missing out a color in the sequence or backtracking). When the pigeon pecks the five keys in the correct order, it gets the grain. Any mistake and it has to sit in darkness for a moment before it gets another chance.

Though they didn't find it easy, Terrace's pigeons ultimately learned this task and would peck the colored keys in the required order. The real surprise came when the pigeons were tested on pairs of colors selected from the original series of five. For example, Terrace's team might just light up the green and yellow keys. Would the pigeon know to peck green first and then yellow, as it had when they were part of a series of five lights that had to be pecked in blue-green-yellow-red-violet order? Most of the time, the pigeons just didn't have a clue. Although a pigeon would happily peck green followed by yellow if it had seen all five colors laid out before it, it had no idea what to do when green and yellow were the only options. Only if the first color in the series (red) or the last (violet) was lit up could the pigeons choose accurately.

But Terrace and his collaborators found a substantial difference when they presented the same problem to monkeys. For the monkeys it was nowhere near as difficult to solve. Not only could the monkeys learn to tap on a specific sequence of five (or more) items when they were all lit up together, but when offered pairs of items selected from inside the larger series, the monkeys readily knew what to do with them too and tapped them in the correct order for the complete series they had first been trained on.

Why should the monkeys and pigeons perform so differently on this task, and, in particular, why should the pigeons have so much difficulty with it when they were just fine on the similar-seeming transitive inference task?

The suggestion that Terrace made is that the pigeons have no way of comprehending the whole series of colors that they need to peck as a series. Instead of somehow representing the whole series to themselves, they solve the task by following simple rules. Perhaps the first rule the pigeon learns is, "When the stimulus lights come on, peck the blue light." Once it has followed that rule, the next rule that comes into play is, "When you have pecked the blue light, peck the green one." This is followed by "When you have pecked green, peck yellow," and so on down through the series. A mechanism of this type would be sufficient to solve the problem when all five lights are on at once. But it could also explain the difficulties the pigeons have when shown pairs of colors selected from the middle of the series. Confronted with just green and yellow (the second and third colors in the series), the pigeon would try and activate Rule 1 ("When the stimulus lights come on, peck the blue light"), but the bird would be stuck, because the blue light was not on. Unable to activate the first rule, Terrace suggests, the pigeons had no way of moving on to the rules that did apply to the colors in the middle of the series, such as "When you have pecked the blue light, peck the green one."

The transitive inference problem can be solved successfully by the application of simple rules, and so pigeons, rats, monkeys, apes, and humans all reach the correct deductive conclusion. The capacity for transitive inference belongs to the filling in the similarity sandwich—it is something that many species share. But Terrace's similar-seeming series task shows the limitations of reasoning with simple rules: pigeons learn the original task but fail to reason deductively when confronted with pairs of items selected from within the original training series. Here we seem to be getting into the bread on top of the similarity sandwich: different species approach this problem differently. Humans, and sometimes monkeys, can do what a pigeon cannot.

LET ME COUNT THE WAYS

Once a child can reason about series of objects he is ready to start reasoning mathematically. According to Piaget, a basic understanding of number starts to kick in during the concrete-operational stage of intellectual development. By age eight most children can grasp basic number concepts and can add, subtract, and deal with simple multiplication and division. Research on the number sense of animals is still very patchy, but the evidence so far indicates that many species have some sense of number, and a few show hints of being able to reason with numbers.

Otto Koehler (who, despite his similar name—Köhler is sometimes spelled Koehler—is no relative of Wolfgang Köhler) was professor of psychology at the University of Danzig (now Gdansk in Poland) in the interwar years. He became fascinated with animal behavior in 1908 while sitting in the same zoology class at the University of Freiburg as Karl von Frisch (he of the honeybee dance communication system; see chapter 2). Where his namesake, Wolfgang K., was able to profit from his experiences in the First World War and got out of Germany before the Second broke out, Otto K. was taken prisoner while working as a nurse in Palestine in the First World War and lost everything in the Second. His laboratory and records were destroyed in the last year of the war, and he was lucky to escape the Russian advance with just his violin and two suitcases. Sympathy with his plight is tempered, however, by the knowledge that he was a card-carrying member of the Nazi party.

In happier times Koehler's group had investigated the sense of number in a varied set of birds: jackdaws, crows, budgerigars, ravens, magpies, and pigeons were favored subjects. By training the birds to choose between containers with different numbers of grains glued to their lids, Koehler and his students demonstrated that these species had a sense of different quantities, even if they didn't have names for numbers. Koehler's group used a simple system of carrot and stick. His students offered the birds choices between containers with different numbers of food grains glued to their lids. If the bird chose the container with the correct number of grains on its lid, the food it found in that container was its reward. If the bird chose incorrectly, it found no food reward, and, if necessary, it was shooed away from the container verbally, by hand, or, in a couple of recalcitrant cases, with something akin to a fly swatter.

Koehler was aware of a problem in his experimental design. The birds may have just been choosing on the basis of the area of the lid covered by grains and not the number of grains as such. The larger the number of food grains placed on a container lid, the more of the area of the lid was obscured. To control for this problem, Koehler's students repeated their experiments with lumps of plasticine of different sizes. Because the plasticine lumps varied in size, the area of the lid obscured by plasticine was unrelated to the number of plasticine lumps. Nonetheless, the birds were still able to choose on the basis of the number of items on the lid.

With these methods Koehler and his group were the first to demonstrate both that animals have a sense of the difference between larger and smaller quantities (*relative number*) and that they can comprehend that number is a quality that groups of different items can have in common. This quality that a certain number of ants shares with the same number of elephants is known as *absolute number*.

In the subsequent seventy years of research some sense of relative and absolute number has been demonstrated in a variety of mammals and birds by researchers around the world. In most studies the animals' sense of number doesn't seem to reach beyond seven. Interestingly, people are also unable to count beyond seven if the objects they are to count are flashed in front of them faster than they can count with words. Some success with larger numbers was achieved in a study where rats were trained to press a lever a fixed number of times to get a reward. The rats could press the lever up to fifty times to get rewarded, but they were not very accurate.

This kind of sensitivity to quantity, though interesting in itself, is not really reasoning. These animals are not using numbers to deduce true consequences, or even adapting their behavior to solve problems. They were just counting. What evidence is there for numerical reasoning?

Here again, Herbert Terrace of Columbia University in New York—this time with his then student Elizabeth Brannon—came to the fore. Brannon and Terrace carried out an interesting study of numerical reasoning by training each of their macaque monkeys to touch patterns that appeared on a computer screen. Macaques are hardy little Asian monkeys that have long been popular in psychological experiments. The patterns that Brannon and Terrace showed their monkeys contained different numbers of objects: one pattern might be made up of three circles, another of a line diagram of a car; yet another might include three different items, such as a plus sign, a sketch of a strawberry, and a shape a little like a pear. These objects varied in color and were presented on square backgrounds of different colors. For their study, the contents of each square were unimportant; all that mattered was the number of objects on each background. The monkeys had to touch the squares that appeared on the screen in the order of the number of objects they contained (some monkeys were trained to touch the squares in ascending order of their numerosities, others in descending order.) The monkeys readily learned to select quantities from one to four in ascending or descending order, even though a large variety of different patterns was used throughout training.

Brannon and Terrace recognized that these monkeys might not have any understanding of number; they might just be learning to a touch series of objects in an arbitrary order, as monkeys trained by Terrace's group had done many times before. The results they obtained might have nothing to do with counting per se but just indicate an animal's memory for arbitrary symbols and how to order them.

In a follow-up experiment to explore this possibility, Brannon and Terrace presented their macaques with just two sets of items on the screen at the same time. A correct response now was one to the larger number of objects. On some tests the monkeys were shown numbers of objects in the one-to-four range that they were familiar with from earlier training. But on other tests the monkeys were confronted with patterns containing from five to nine items. The macaques had never seen objects in this range before, and their choices from among these novel numbers were never rewarded; they therefore serve as a pure test of whether the monkeys had really abstracted the rule involved in counting objects. All three monkeys performed at very high levels in choosing numbers in the familiar one-to-four range. Choice in the novel five-to-nine range was less reliable, but two of the three monkeys were able to do better than chance. Thus there is the suggestion here that the macaque monkeys were able to deduce from their experience with quantities from one to four what to do with numbers of items in the five-to-nine range.

Numbers are useful to us for more than just counting objects; they can also be used to add and subtract quantities. The only suggestion that an animal can use numbers in this way comes from one of Sally Boysen's chimps, Sheba, at Ohio State University. We saw above, in considering Limongelli and Visalberghi's trap tube task, that Sheba is a highly gifted animal. Boysen and her colleagues first trained Sheba to select cards with Arabic numerals on them to match the number of objects presented on a tray. Thus, Sheba might be shown three candies and asked to select the correct card to match the number of candies. Three alternative cards were offered with "1," "2," or "3" on them. Once she had mastered this, the task was reversed. Sheba was shown a card with a numeral on it and required to pick the tray of candies with the matching number of items.

Sheba's pièce de resistance tested her arithmetical reasoning skills by building on what she had learned about numerals. She was trained to collect oranges from two different places. Sheba then had to report how many oranges she had found by picking the appropriate numeral from a selection. The total number of oranges could not exceed four—the largest number Sheba was familiar with. Sheba quickly learned, for example, that if there was one orange in one place, and two more in another, that the numeral she must select was "3." Finally Boysen tested Sheba by replacing the oranges with cards with Arabic numerals on them. Sheba went round the room collecting numerals from two locations and then selected the number that matched the total she had found. She was able to transfer her experience with oranges immediately to the numerals and was highly successful, performing correctly on three-quarters of the trials given to her.

Sheba's performance may be the first demonstration of arithmetical reasoning in a nonhuman. But before we pop the champagne, it is worth considering a limitation to this experiment, first pointed out by science journalist Stephen Budiansky. Sheba was only tested with totals up to four. The problem with using such small numbers is that there are not many different ways that two positive integers can be added together to give a total not larger than four. Sheba could just learn a limited set of rules. For example, "If you find one orange, pick 1"; "If you find one orange twice, or two oranges in one place and no others, pick 2"; "If you uncover one orange in one place and two in another or three in one place, pick 3." Even a total of four can only be arrived at in three ways: zero and four, one plus three, and two plus two. It is possible that Sheba just mastered this problem as a small set of rules that had to be rote-learned. Only as the numbers that a chimp (or a human child) can cope with gets larger does a solution to arithmetic problems based on rote learning of rules become so cumbersome that it can be ruled out as a possible explanation for success.

My feeling is that, at present, arithmetical reasoning has not been unequivocally demonstrated in any animal species. Sheba may be capable of it, but we can't be confident. There is certainly evidence for some protomathematical skills, like an appreciation of quantities and the ability to order them, in a range of birds and mammals, but we are still waiting for evidence of true mathematical reasoning in animals.

PATTERNS OF REASON

What do we find when we consider reasoning in animals? Sadly, we find that only a handful of species have ever been tested and on just a tiny number of possible forms of reasoning. Yet even though the pickings are meager, I see a clear pattern emerging.

Let's first look at reasoning in the broader sense of adapting behavior to solve problems. Adaptive behavior is extremely widespread. In fact (though this statement may seem bold given how few species have been studied), I believe that all animals are capable of some kind of adaptive behavior—from the wasp sniffing out plants that have been bitten by caterpillars, through the dog that attends to any word or action that predicts food or a stroll in the park, to the infant who knows her cries of distress will bring concerned parents running even at 3 A.M. In this broader sense reasoning belongs in the squidgy filling of the similarity sandwich: it is something that all species share.

But in the more limited sense of reasoning, as a Sherlock-Holmesian ability to deduce true conclusions from valid premises, the picture is much muddier. There is, nonetheless, a moral that can be drawn from the mixed results of various experiments: many species can succeed on some astonishingly complex-seeming problems by the application of pretty simple rules of thumb. Conversely, these species may fail on tasks that seem to us no more complex than the first kind, because their rules of thumb let them down. I'm thinking of the pigeons that could solve the transitive inference problem (reasoning that if blue is better than green and green is better than yellow, then blue must be better than yellow) but failed on Terrace's task, where a very similar question was put to them but in a slightly different way.

How unlike our own logic, with its its apparently infinite flexibility. We bring our reasoning to bear on any problem that comes our way; we are not just applying rules mechanically. Our ability to reason might feel like that, but it turns out that this is something of an illusion. Consider these two problems.

First, imagine you are given four cards and told to test the rule that a card with a vowel on one side must have an even number on the other side. Let's say the cards in front of you show an *E*, a *K*, a 7, and a 4. Which would you turn over? Most people find this a very difficult problem. Most turn over the *E* and some also turn over the 4. And yet the 4 can tell you nothing: Who cares what's on the other side of an even number? the rule being tested does not say that the flip side of an even-numbered card can not be a consonant, only that the flip side of a card with a vowel cannot be an odd number. So you would learn nothing by turning over the 4. The correct answer is to turn over the *E* (see if the vowel has an even number on its reverse) and the 7 (check that there's no vowel there). Only about 5 percent even of the college-educated population give the right answer on this one. It's a tough logical nut to crack.

Now consider this problem. Imagine that you are shown four people and told to test the rule that a person must be over the age of twenty-one to drink beer. One person is drinking Coke; one is drinking beer, the third is twenty-three years old; and the fourth, fifteen. Whom must you check (what they are drinking or what age they are) to ensure that the rule is being followed? Here nobody has any trouble. We don't care what the twenty-three-year-old drinks, nor what age the Coke drinker is, but we do need to check the age of the beer drinker and the beverage of the fifteen-year old. Nothing could be simpler. Hardly anybody gets this one wrong.

And yet logically these are absolutely identical problems. There is no difference in the type of reasoning required to solve these two puzzles. Why the big difference in performance?

Maybe our wonderful rationality isn't so open-ended after all; maybe we also are constrained to reason best about certain types of problems. Problems that involve detecting people who cheat against the rules of society seem very important to us, social animals that we are, and consequently their solution feels quite effortless. Our reasoning engines are focused on puzzles of this type. Problems with the same degree of logical complexity, set in terms of arbitrary symbols, strike us as far more difficult. So one thing to consider in comparing ourselves to other species is that the way our reasoning abilities feel to us is not necessarily the way they would look to another species that wanted to study us.

So should we consider deductive reasoning, even in the narrower Sherlock-Holmesian sense, as part of the "all species do this" filling of the similarity sandwich or as part of the "only humans do this" bread on top? My reasoning tells me that if pigeons do it, then it's not unique to humans. But I do see a difference of degree here that is not trivial. Although, as I've said, our reasoning isn't as open-ended as we'd like to imagine, it would be romantic dreaming to pretend that our deductive abilities don't stretch further than those of most other species.

No animal but *Homo sapiens* could figure out what Wolfgang Köhler had been up to on Tenerife all those years ago. No other animal would set another species problems in order to comprehend the limits of its reasoning abilities. So there are both qualitative and quantitative differences between our reasoning powers and those of other animals. But when we break reasoning problems down into their component pieces, we see that other species do succeed on some parts of the reasoning puzzle. The difference between us and them is not so hard and fast. Let's turn the similarity sandwich into a blueberry muffin. There are bits of similarity between us and them. These bits of blueberry catch our attention and spark our interest. But there is still more muffin than blueberry: a lot that we do that they can't.

FURTHER READING

The Hunting Wasps, by J. Henri Fabre (Hodder and Stoughton, 1919). A tremendously readable personal account of simple yet powerful experiments on wasps that any modern reader could profitably emulate. Unfortunately, the English terminology used in the translation is rather obscure in parts, but Fabre's enthusiasm for his animals more than makes up for any difficulty in the vocabulary.

The Mentality of Apes, by Wolfgang Köhler, 2d ed. (Harcourt, Brace, 1927). Köhler's own account of what he did on Tenerife in the first six months of 1914 is still a good read. The illustrations are fascinating too.

A Whisper of Espionage, by Ronald Ley (Avery, 1990). Ley's account of how he uncovered Köhler's role in German espionage on Tenerife in World War I makes intriguing reading.

Inside the Animal Mind: Who's the Birdbrain? <www.pbs.org/wnet/nature/animalmind/intelligence.html>. This web site, the home page of a TV documentary, includes video of a chimpanzee trying, like Köhler's chimp Sultan, to reach a banana by climbing on boxes.

What Is It Like to Be a Bat?

When I decided to borrow the title for this chapter from a famous paper by New York University philosopher Thomas Nagel, I didn't know that British novelist David Lodge, in his novel *Thinks* . . . had already asked what other writers might compose under the title "What is it like to be a bat?" One M*rt*n Am*s, presumably the author of *Money* and *Success*, finds a bat's life to be obsessed with sex and crap. For Irv*ne W*lsh (author of *Trainspotting*), a vampire bat's incessant search for fresh blood is like a Scots junkie's search for heroin—complete with the risk of HIV. S*lm*n R*shd**'s bat is preoccupied with caste and rank, and tormented by his self-awareness. Finally S*m**l B*ck*tt's bat leads a squalid stripped-down existence, trapped in a world of darkness. All of these parodies serve to emphasize the philosopher Nagel's point: we cannot know what it is like to be a bat.

Nagel is convinced that bats are conscious (he doesn't say how he knows this). Consciousness—like love, sex, and football—is one of those words that can mean very different things to different people at different times. But Nagel spells out what he means by consciousness: it means that there is something that it is like to be a bat. Isn't it funny how philosophers can use perfectly harmless little words and end up saying something so apparently incomprehensible? In fact, Nagel's point is not so hard to grasp as it seems: he means simply that it feels like something to be a bat. A bat experiences something. Just as you or I experience sunshine, happiness, and rain, so too, according to Nagel, bats experience things.

That's Nagel's starting point: bats have subjective experiences. But what these experiences feel like to a bat, that is impossible for us to know. We could guess what it might feel like for us if we were bats, but, as Lodge's pastiches of Am*s, W*lsh, R*shd**, and B*ck*tt show us in the garish colors of satire, our guesses would be reflections of our human concerns, not the accounts bats would give of themselves. Those accounts must forever be hidden from us.

Though I don't understand how he comes to the conclusion that bats are conscious, I do think that Nagel's definition of consciousness is a good one. Many commentators on animal consciousness get sidetracked into thinking that there is something an animal might do that would indicate whether it is conscious or not. Some authors think that reasoning, such as we discussed in chapter 3, is evidence of consciousness. Other commentators think that certain kinds of responsiveness to what others know about a situation (as we shall discuss in chapter 5) could be evidence of consciousness. I agree with Nagel that consciousness is something about our experience of the world: a being (human or not) could be completely inactive, seem like a vegetable to an outside observer, and yet still be conscious. On the other hand a being could be ever so clever, could be reasoning, could be acting on the basis of what others know, and yet still not be conscious.

I am not here going to attempt what the philosopher proves is impossible and the novelist entertainingly ridicules. I am not going to try and describe how the world feels to a bat. What I want to do is to look at what we can and do know about bats. The knowable is terrific stuff—rich consolation for Nagel's unknowable bat consciousness. These beasts are so far from the mammalian mainstream. They fly in the dark with unerring precision. They drink blood (some of them, at least), and these blood-drinking bats are some of the few animals proven to behave altruistically toward anyone outside close kin. Bats' peculiar habits have given them a special status in human society; in the West, a very negative one. Other peoples have been puzzled by bats too, but their puzzlement has not translated into demonization the way ours has.

LOVE THAT BAT

Why do we hate bats so? "Bats out of hell,"—we say. Why such vehemence, such aggression against harmless and fascinating animals? Bats are among the most intriguing and diverse of mammals. The bumblebee bat of Thailand, weighing in at less than one-tenth of an ounce, is the smallest mammal in the world; the largest bat is the giant flying fox of India, with a wingspan of two yards and a weight of two and a half lbs.

Of the 4,200 mammalian species alive today, almost 1,000 are bats. If things don't appear that way in your neck of the woods, that might be because you aren't in the tropics (only 12 percent of bat species live in temperate regions), or you don't get out much at night. And yet, although bats are varied, they are not easily confused with other mammals. Why is that? Well, most obviously, bats fly, and other mammals don't. But that is not all that makes bats unique. Bats live much longer than other mammals of their size (up to twenty years or more); they develop slower and have longer periods of maternal care. And mother bats have fewer offspring than other small mammals (usually just one or two). Although they fly, bats are not birds but mammals. They have far fewer offspring than birds, leathery, not feathery wings, and furry bodies. Look at the skeleton of a bat, and there in the wings you will see bones similar to those of our own hands: a thumb, an index finger, and three other fingers. The scientific names for the two suborders of bats, the Megachiroptera (big hand-wings) and Microchiroptera (small hand-wings) reflect this.

The earliest bats in the fossil record were full-fledged flying mammals in the tropical forests of the Eocene around fifty million years ago, surprisingly similar to modern bats. It is a fair assump-

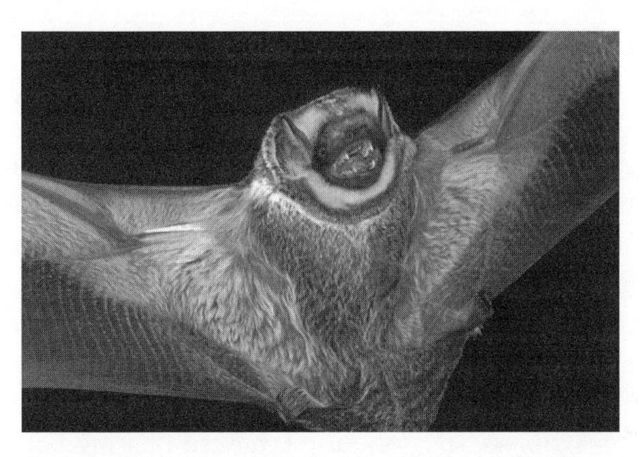

The awesomely beautiful hoary bat. In nature its head is rimmed with gold and its wings are highlighted in red. (© 2002 www.arttoday.com)

tion that bats evolved from gliding mammals, though exactly what sort of mammal remains a mystery.

Megachiroptera are the medium-to-large fruit-eating bats of the Old World tropics, also known as flying foxes, fruit bats, or megabats. These bats rely mainly on vision for navigation (no, bats are not blind). Some of them do echolocate, but only with rather low-frequency sounds that they make with their tongues. (More on echolocation later.) Megabats feed mainly on fruit, flowers, and pollen; they roost in trees and sometimes caves—often in large flocks.

The other suborder of bats is the Microchiroptera, or microbats. These bats are found throughout the world and rely extensively on high-frequency echolocation for navigation. They fly with greater maneuverability than megabats, and many of them catch insects on the wing. Other prey for microbats includes insects and small animals (e.g., frogs), fish, birds, fruit, nectar, and pollen (bats that feed on flowers hover like hummingbirds), pepper plants, and—yes—the blood of living creatures. Each species of bat is beautifully (I almost said "cunningly") adapted to catch its prey. The fishing bat, *Noctilio leporinus*, for example, captures small fish as it flies over a river's surface by dragging its long legs through the water. It detects telltale ripples on the water's surface, and its very long and wide wings ensure maximum lift at low speeds to rip the fish out of the water. Its ugly face has given this Mexican bat the common name "bulldog bat." Frog-eating bats recognize the croaks of their favorite food.

The heads of different bat species are gloriously varied—some cute, some quite cruel-looking—as their common names suggest: dog-faced fruit bat, Hammer-headed fruit bat, tube-nosed fruit bat. The weird characters in Mayan art that the fantasist Erich von Däniken insisted, in *Chariots of the Gods*, were characters from outer space are actually bat gods. The strange headgear, faintly reminiscent of an astronaut's helmet, is a stylized representation of the nose of *Artibeus,* the common large fruit-eating bat of Central America.

Fruit bats pollinate many species of plants, including some important crops like avocados, bananas, cashew nuts, figs, mangos, and peaches. The only other economic value of bats lies in their shit. Bat guano used to be harvested in many of the warmer countries of the world and made an excellent fertilizer. There was even a fad in Texas in the early part of the twentieth century for having municipal bat roosts where bats would be encouraged to stay. The hope was that insectivorous bat species would keep the mosquito numbers down and produce dung for good fertilizer. Unfortunately, it proved difficult to get these bats to stay in the roosts prepared for them.

SEE ME, HEAR ME, TOUCH ME . . .

So how do bats find their way in the dark? It was the discovery of the answer to this question that prompted philosopher Thomas Nagel to ask what it might be like to be a bat. What a question this is: In a world of total blank nothingness, how do bats orient swiftly in three dimensions? And once we have the answer, we have a new question: What kind of a world might that be, perceived in this utterly alien way?

Lazzaro Spallanzani, bishop of Padua, Italy, in the late eighteenth century, took the first steps toward an understanding of bat navigation. The bishop was curious about the possible "sixth

A detail from the Madrid Codex of Mayan art as selected by Erich von Däniken. According to Däniken, these Mayan artworks "depict the whole arsenal of space-flight paraphernalia: supply systems, helmets with transmitters, an observer in satellite, and oxygen apparatus." Aside from the curiosity that these prehistoric space travelers used Apollo-era technology, in reality these are images of Mayan bat gods with the heads of bats and the torsos of men. (Erich von Däniken, *In Search of Ancient Gods: My Pictorial Evidence for the Impossible*, 1973, Econ Verlag)

sense" that enabled owls and bats to fly at night. I don't know where the cleric learned his scientific method, but he was good at it. He hung thin wires from the ceiling of a windowless room. At the bottom of each wire Spallanzani tied a tiny bell so that he could hear if one of his flying animals hit a wire. Did these night flyers have a sixth sense? With the owls all Spallanzani had to do was blow out the candle and they refused to fly. Spallanzani concluded quite correctly that owls use very sensitive vision to guide their flight. But with the bats the situation was quite different. The bats would happily dodge the wires, not ringing a single bell, regardless of how dark it was. Being a good, if not humane, experimentalist, the bishop then blinded the bats with a red-hot poker—there was no difference to their flight patterns. He even cut their eyes out of their heads. The bats flew on, regardless.

But then Spallanzani placed little brass tubes in the bats' ear canals. So long as the tubes remained open, the bats were able to dodge the wires without trouble. As soon as the bishop closed the little tubes, so that the bats could no longer hear, the bells on the ends of the wires started ringing. Clearly the bats needed their hearing to fly in the dark. Spallanzani had made an important discovery: bats need their ears to navigate in the dark. What sounds were they relying on? He listened carefully but could hear nothing. Progress had been made, but the mystery remained: how could the bats be using their ears to "see" in the dark?

Unfortunately, the generations of scientists who followed Spallanzani were unable to conceive of a system of navigation based on hearing. Baron Georges Cuvier, an influential French contemporary of Spallanzani's, insisted that bats use an ultrasensitive system of touch to "feel" their way through the night sky. Though Cuvier had not a jot of evidence for his theory, his account held sway well into the twentieth century.

In 1912 the American-turned-British-subject Sir Hiram Maxim, inventor of the machine gun, was the first to propose that bats might navigate using echoes. Doubtless unfairly and without a jot of evidence, I picture Sir Hiram as a rather overweight and jovial man on the pattern of nineteenth-century American entrepreneurs in BBC costume dramas. Maxim suggested that bats emit sounds too deep in tone to be heard by the human ear. He was close: we now know that we do not hear bat sounds because they are too high, not too low, in pitch for our ears. Maxim suggested that the reflections of these sounds were picked up by highly touch-sensitive pads on the wings that Cuvier had (erroneously) claimed to have found.

Maxim was so close. If he had thought of the possibility of echolocation, why did he not look at the enormous ears of these animals? It seems that Cuvier's reputation was just so much stronger than Spallanzani's that, even though simple experiment would have shown that ears and not touch receptors on the skin were essential to flight, Maxim stuck with Cuvier's account.

Echolocation may seem alien to us humans, but we are not entirely incapable of it. There is a game three or four people can play that shows our surprising ability to locate things with echoes. One blindfolded person is the research subject. One or two others hold a large piece of card a few feet away from the blindfolded person. The final person in the group measures the distance between the card and the subject. If the blindfolded person is allowed to make different sounds in the direction of the cardboard and then asked to guess how far away it is, she may do surprisingly well. Most people can at least tell if the card has been moved closer or further away from them and, if given feedback after each guess, gradually make better estimates of its distance. At the very least, any blindfolded person can tell easily if she is in a small or large room. Many blind people rely on echoes to avoid bumping into things but, interestingly, are often not aware that they are doing so.

Only in the 1930s, through the compelling experimentation of Donald Griffin, was the mystery of the bat's echolocation system unraveled. Though I have difficulties with Griffin's views on animal consciousness (I went over our differences in chapters 1 and 2 and won't repeat them here), I have enormous respect for the research he carried out—starting while still just an undergraduate at Harvard. Griffin enlisted the help of a physics professor, George W. Pierce, to demonstrate that bats emit sound at frequencies higher than the human ear can hear (known as ultrasound) and use the echoes from these calls to locate objects in space. Then, with the help of Robert Galambos, Griffin carried out a series of experiments testing bats with blindfolds, earmuffs, and gags round their mouths—essentially replicating Spallanzani's work of 150 years earlier, and demonstrating definitively that bats navigate through the use of ultrasonic echoes.

To understand bat echolocation, we must first appreciate that the sounds we hear come to our ears as waves of pressure through the air. Air can be compressed (bunched up) or rarefied (allowed to ease out), and these compressions and rarefactions pass through air like waves on the surface of an ocean. As these waves lap at our eardrums, we hear sounds.

Sound waves travel at around 770 miles per hour, and what we hear as different tones are actually waves of different frequencies. The standard A that orchestras tune to, for example, has a frequency of 440 oscillations per second; this means that every second 440 waves of compressed air beat against our eardrums. Technically, one cycle of compressed and rarefied air every second is known as a Hertz (Hz). So the concert A is 440 Hz. The highest tones a healthy young human can hear are around 20,000 Hz (20 kilohertz or kHz), which sounds like the highest-pitched whistle.

Only a few sources of sound produce just one frequency: tuning forks and violins, for example, produce fairly "pure tones," as musicians call them. Scientists know these as constant-frequency sounds. Most sources of sound mix together many frequencies; the most common are harmonics at the octaves (what a musician calls an octave a physicist recognizes as a doubling of frequency), but many combinations of frequency are possible. Mixed, muddled, and changing frequencies are known as frequency-modulated sounds. Vowels are closer to constant frequency sounds; say "ah" and you are probably close to producing one or just a few frequencies. Consonants are very frequency-modulated; say "tut" and you are emitting a frequency-modulated sound.

Bats use both constant-frequency and frequency-modulated sounds in echolocation. They also use very high-pitched sounds. Only the lowest of bats' sounds dip into the range of frequencies audible to humans (the Dutch zoologist Sven Dijkgraf, working in Nazi-occupied Holland with no technology but his own very sensitive ears, figured independently of Griffin that bats were using high-frequency sound to echolocate).

Just like the waves created by plucking a long clothesline, the longer a sound wave is (the further from trough to trough and peak to peak), the slower it oscillates and the lower its frequency. Deep tones are made up of longer waves than those in high-pitched sound. Thus, the advantage to bats of using high-pitched sound to echolocate is that these sounds are carried on smaller waves and consequently permit the localization of smaller objects. This is a fact of physics. I remember as a kid noticing how waves in the seas around the Isle of Wight would bounce back from large objects like boats but would slip past thin things like the posts holding up Shanklin Pier without apparently recognizing any barrier. Waves of any kind are reflected only from objects larger than themselves; objects smaller than the length of the wave present no barrier to them. What is true for waves in the waters around the Isle of Wight is equally true for sound waves.

A musician who tried to echolocate with a tuning fork vibrating at 440 Hz would only be able to detect objects larger than the length of those waves—about two and a half feet. Even the highest note on a piano could only be used to detect objects bigger than about four inches. This would be no way to hunt moths. Bats' hunting calls vary from around 20 kHz to over 100 kHz—many octaves above the range of human hearing. These high frequencies produce waves whose lengths are measured in hundredths of an inch. This ensures that no objects of significance can slip past them. In experimental tests, bats can locate objects smaller than one-twentieth of an inch in diameter.

To catch an insect on the wing, a bat, like a fighter pilot trying to destroy an enemy aircraft, would like to know the range, size, rate of approach, and bearing both in the vertical (elevation) and horizontal (eccentricity) dimensions of its prey. All of this information can be gleaned through the bat's sonar system.

To find distance information, all a bat needs to do is emit a sound and time how long it takes for the echo to return. This sounds simple, but sound travels at 770 mph, and therefore the time intervals involved are often very short. As the bat comes in for the kill, for example, at a distance of seven feet there are just twelve-thousandths of a second between emitting a sound and hearing its echo. The bat brain has areas dedicated exclusively to registering these brief time intervals.

So bats get distance information by timing echoes. How do they pick up object size? As an object gets bigger, it will reflect more sound and consequently produce a louder echo. But this could be confusing. A small object up close may produce just as loud an echo as a larger object further away. The bat, however, already knows how far away the object is (from the time it took the echo to return), and so it can take this into account when listening to the loudness of the echo and thereby calculate the object's size.

Bats also glean information from their sonar system about the rate at which prey is approaching. To understand how they do this, we need to consider the experiments of the Austrian physicist Christian Johann Doppler in the mid-nineteenth century. Doppler set a group of musicians on an open carriage on a train and had them pulled past another group of musicians on a station platform. The musicians on the train were instructed to play one continuous note. The musicians on the platform had perfect pitch and recorded how the note from the train appeared to them to change in pitch as the train went past. Just as any child listening to a passing fire engine can tell, a note goes up in pitch as the moving object approaches and down again as it recedes, and the rate at which the pitch changes depends on how fast the fire engine is moving. This is because, as the object moves toward the observer, the motion of the sound source adds to the speed of the sound waves through the air: the sound waves become "bunched up," and this raises their frequency and consequently their pitch. As the sound-producing object recedes, the motion of the sound source subtracts from the speed of the sound through the air: the sound waves are spread out, and thus their frequency and pitch go down.

This phenomenon, now known as the Doppler shift, offers both a problem and an opportunity to echolocating bats. The problem lies in the effect of the bat's own movement on the pitch of the echoes it hears. How is a bat to keep track of its echoes if they vary in pitch due to its own flight? But the Doppler shift also creates an opportunity for a bat to estimate how fast prey is moving toward it by attending to the change in pitch of the echo.

To get over the problem created by their own movement, many bats have a frequency-modulated component to their calls. If a call contains many frequencies to start with, it hardly matters if the Doppler shift alters the frequencies in the echo; it is still possible to extract range information from the time that elapses between call and echo, and size information from echo loudness.

To exploit the possibilities inherent in the Doppler shift for detecting prey movement, many bats also have a constant-frequency component to their call. Obviously bats hear their own calls as they emit them (though we don't hear them, bats' calls are actually extremely loud—so loud that bats' ears come with a special dampening mechanism that is activated when the bat emits a call and prevents ear damage). If the bat's call is of a pure frequency, then it can compare the frequency it sends out to the frequency of the echo that comes back. From this comparison the bat can pick up the Doppler shift and thereby gauge the approach velocity of its prey. In fact, the way bats do this is by changing the tone of the ultrasound signal that they send out in order to keep the returning echo always at roughly the same frequency. The reason for this is that a great deal of neural machinery is required to process the returning echoes, so the brain can function much more efficiently if the return echo can be kept within a narrow band of frequencies than if it varied widely.

Interestingly, bats only perform this Doppler correction for echoes that indicate an object is getting closer. If the object appears to be receding, no correction for the Doppler shift is attempted. The reason for this is simple. There are only two ways that an object could appear to recede. Either the bat must be flying backwards, which is impossible, or the prey object is flying faster than the bat—in which case it might as well be ignored.

Bats identify the left/right position (eccentricity) of an object the same way we do, by having two ears. The sound from any object not dead center is slightly louder at one ear than the other, and this gives away its left/right position. (This is also the principle behind stereo sound systems.) How bats identify up/down position (elevation) is less well understood. One possibility is that the elaborate and often very odd shapes of bat ears ensure that echoes coming from above and beneath them have different tonal qualities, the way that the same note played on different instruments has different qualities, and this might make it possible to gauge an object's elevation.

By the late 1950s Griffin and his associates had demonstrated that bats could locate, track, and capture flying moths, small flies, and mosquitoes using their sonar system. At first Griffin's experiments were not that different from Spallanzani's two centuries earlier, except that Griffin had a microphone sensitive to higher frequencies than the human ear can hear (so he could pick up the sounds the bats were making); and he "blinded" his bats harmlessly with blindfolds instead of a red-hot poker. But as research into the hitherto unknown sonar world of the bat intensified, bats were brought into soundproofed rooms rather like recording studios. In these carefully constructed environments, a bat might be trained to stand on a perch so that its vocalizations could be recorded. In this way the pattern and intensity of bat sounds were gradually understood. Or the bat might be the audience as experimenters played different sounds to it. These sounds might be recordings of moth noises, played back from different directions to see how the bat responded to them or modified in various ways to see what components of the sound were important for bat orientation and prey capture. Most recently, tiny microphones with radio transmitters have been developed that can be strapped onto the bat's back. In this way, rather like pop stars with cordless mikes, bats' calls can be recorded while the bat is on the wing.

Griffin's ultrasonic microphones identified that bats emit ultrasonic "chirps" lasting about 15 msec (thousandths of a second) when searching for prey. These change to a continuous "buzz" when the bat goes onto the attack or avoids an obstacle. The echolocating abilities of some species of bat are so finely tuned that they can perceive the texture of a surface, or the change in wing beat of a moth.

But the evolutionary arms race of predator and prey is rarely a one-sided battle. In the early 1960s Kenneth Roeder and Asher

Treat at Tufts University demonstrated that the noctuid moth, a species regularly preyed upon by North American bats, has a specialized ear, known as a tympanic organ, tuned for the detection of bat ultrasound. Of the sixty-eight species of moth preyed on by bats in India, all but three are sensitive to bat ultrasounds. Shortly after it was discovered that moths could hear bat signals, it was also found that some moths actually emit ultrasound clicks of their own. These moths wait till they hear the bat switch from its searching chirps to its attacking buzz before making their own ultrasounds. These effectively stop the bat's attack, though exactly how is a matter of ongoing debate. One possibility is that the moth's clicks interfere with the bat's ultrasound. Another speculation is that the bat hears the moth's clicks and interprets them as a signal that the moth is unpalatable and best left alone.

Bat echolocation is in my view one of the most astonishing discoveries made about any animal's world in the last fifty years. And yet, although echolocation seems so exciting compared to boring, old seeing-by-light, there are drawbacks to using sounds to find things. For one, animals that use light to navigate can rely on the free supply of light energy from the sun, while an echolocator has to provide sound itself. This is not a trivial energy demand. The horseshoe bat emits around 400,000 calls a night, and each call is as loud as a rock band or a plane landing as heard from the runway. It is just as well that the energy in these calls is outside the range picked up by the human ear; otherwise it wouldn't just be the bells in the belfry that would sound deafeningly loud. To generate enough energy in their calls, very high air pressures are needed, only slightly lower than blood pressure (any higher and the bat would faint every time it tried to call). Bats have solved the energy problem, however, and no energy is required for ultrasonic calling beyond that demanded for flying. The bat only calls as it contracts its chest muscles to flap its wings, this forces air out of the lungs as a costless side effect. As the air is forced through the larynx (a voice box in the throat), sound is produced. Very few nonflying mammals try to echolocate, because if you're not already flapping wings energetically, the extra effort of emitting sound calls loud enough to be of any use would be just too great.

But even with its 400,000 calls a night, the bat is still "in the dark," without any information about its surroundings, for more than half of each eleven-hour nighttime shift. Each call can only be very brief; the bat gets more like a camera flash view than the illumination a street light might provide. Furthermore, this flashbulb view is also very narrow: the bat only "sees" in the direction it points its sound beam. To "look" around it, the bat would have to emit pulses in several different directions, which would cost extra energy and would also reveal the bat's location to those moths sensitive to ultrasound. And even though the energy required to produce ultrasound is very great, the range of most bat calls is typically only a little over twenty yards or sixty to seventy yards at most. So the bat "sees" only in brief glimpses, only in one direction, and only a few seconds' flight ahead of itself.

This is a restricted world compared to the visual one we are accustomed to, but it is still very rich in its own way. The recent work on bat echolocation shows that this phenomenon has been misnamed: we are not dealing here with a system simply of location but of perception. Echoperception enables bats to "see" the world—not just to avoid obstacles but also to perceive the position, shape, and even texture of objects. Some bats can even "see" the rapidly beating wings of an insect.

It was ultrasonic echolocation that brought bats to the philosopher Nagel's attention. With some effort of imagination, we might be able to fantasize about what it would feel like to fly at night relying on a narrow-beam flashbulb to light the way. But we'd still only be imagining what *we* would experience if we were bats, which, as the philosopher Nagel pointed out and the novelist Lodge accentuated, is simply not the same thing as what the bat experiences. Assuming that the bat experiences anything at all. So we will never be able to satisfy Nagel's and Lodge's curiosity about what it feels like to be a bat, but we can identify how echolocation works and what it achieves for the bat. We know what a bat can perceive, the outline of what its world contains. And this knowledge does not make the world of the bat any less alien.

BAD BATS, GOOD BATS

Though the question of how the world might appear to a bat only piqued people's curiosity in the fifty years since Griffin's discovery of echolocation, the strangeness of bats has puzzled people for thousands of years. Simply being mammals that fly is weird, and flying at night when most birds are asleep is even stranger. Where pigeons/doves could become holy because they commuted between the heavens and the earth (see chapter 4), the bat's penchant for appearing suddenly in the sky at night out of caves and empty houses led to fear, prejudice, and an association with the devil.

The Western tradition is entirely negative about bats. Bats accompany the dead to Hades in Greek mythology. In Leviticus, in Isaiah, and in the apocryphal Book of Baruch, bats are treated as evil, loathsome symbols of darkness. In medieval Europe, imps and devils were regularly shown with batlike leathery wings and pointed ears. This association was so compelling that, when one of the sailors on Captain Cook's voyage to Australia in 1770 came across a large fruit bat, he thought he had seen the devil incarnate.

Materializing out of caves at night is spooky enough, but if just one habit could justify all the bad publicity that bats get, it would have to be sucking blood from living people (though actually, vampires don't suck; they lap). There are three species of vampire bat, all found exclusively in Central and South America (notwithstanding the absurd claims of a literary commentator on Bram Stoker's *Dracula* that people in England had been infected with rabies from vampire bat bites as recently as 1970). Two vampire species feed on birds, but one, the common vampire bat, feeds solely on mammals, nowadays usually cattle. This bat gouges a small depression in the skin of its victim and laps up the blood. It secretes an agent that suppresses clotting so that it can lap for longer. But even so they don't take much, and it doesn't hurt. Although these bats are responsible for a lot of bad bat publicity, they only take about half an ounce of blood in one night—about half their own body weight, and to do so typically takes around twenty-five minutes of gentle lapping. Since human blood donors regularly volunteer a pint in one sitting, giving half an ounce to a bat is hardly going to harm anybody. And the fact that vampire bats only get their teeth into sleeping victims means that those who have been bitten by bats don't notice till the next morning.

Nor is it easy to get a vampire bat to bite a human being. One crazy nineteenth-century explorer traveled round South America with his foot dangling out of his hammock at night in the hope of experiencing the fangs of the vampire—but all in vain. His Scottish traveling companion was more "fortunate":

> On examining his foot, I found the Vampire had tapped his great toe: there was a wound somewhat less than that made by a leech; the blood was still oozing from it; I conjectured he might have lost from ten to twelve ounces of blood [wildly exaggerated]. Whilst examining it, I think I put him into a worse humor by remarking, that an European surgeon would not have been so generous as to have blooded him without making a charge. He looked up in my face, but did not say a word: I saw he was of the opinion that I had better have spared this piece of ill-timed levity.

Rare indeed is the author of a travelogue who succeeds in conveying just what an irritating traveling companion he must have been! Unfortunately, though the simple fact of extracting blood may not do much harm directly, the feeding habits of the common vampire bat can lead to the spread of blood-born disease. Anxiety about the spread of rabies has prompted massive human destruction of vampires in their native climes. But the fear and hatred of vampire bats is out of all proportion to the possible harm they may do. The perception of vampires has a lot more to do with Bram Stoker's masterly 1897 novel *Dracula* than with the facts of vampire life. Vampires turn out to be among nature's most caring and sharing animals, as we shall see in a moment.

Stoker's genius was to take some very ancient and primeval eastern European stories about the *vampir*, *a* person who returns from the dead to feed on the blood of the living, and splice them onto the accounts which by the mid-nineteenth century had reached Europe from Central America about blood-feeding bats that were given the same name. Since losing blood causes death, it was an understandable superstition that drinking blood might restore the dead to life. But meld this belief with old prejudices against bats, with their cloak-like wings and nocturnal flights, and the new knowledge of their blood-sucking habits, and you have a brand so powerful that new Dracula movies appear on average every two to three years.

And yet there are no vampire bats anywhere in Europe. Why not? Because blood is not the most energy-rich of foods: it only has enough calories to sustain a bat who lives in a warmer clime. In the tropics the blood of a human being (about eleven pints) would provide rations to keep a bat happy for a year. In northern Europe (or most of the United States), a bat would have to drink that much blood each month to stay alive. The vampire would need to take in so much blood in each feed that it would be unable to fly away again afterward.

YOU SCRATCH MY BACK . . .

Vampire bats do not have to do anything else to get our attention: just taking blood from the living is enough. But vampires have more going for them than that. I already mentioned that blood is not an especially energy-rich food (which is why we harvest milk from our cattle and not blood). Around 8 percent of vampire bats fail to find a meal on an average night: that means that once every twelve nights a vampire goes out hunting and comes home with an empty belly. For young vampires the risk of not getting a blood meal is much higher, more like once every three nights. Ticks, leeches, and insects that suck blood can go for weeks without a meal. They are cold-blooded and need little energy to stave off starvation. But warm-blooded vampires need a lot more energy to stay alive. Every vampire bat needs to get a meal of blood every night or else it will starve to death in just two days. Here's a problem in probability for you: If 8 percent of bats fail to get a blood meal on an average night, and two nights without food is enough to kill a bat, what is the vampire's average life expectancy? If you said "short," fair enough, because we know that the life expectancy of the vampire bats is actually very long: about eighteen years. So how do they do it? How do they survive so long under such uncertain circumstances? The answer is altruism.

We saw in chapter 2 that it takes special conditions for cooperation among animals to evolve. In honeybees it is the fact that the members of the hive are such very close kin that has made cooperation possible. Where pioneer geneticist J.B.S. Haldane would lay down his life for three brothers or nine cousins, honeybees can happily give theirs for two supersisters. Such is the power of the selfish gene. Helping our kin is as good as helping ourselves—in proportion to the number of genes we have in common with the individual to be helped.

But I also mentioned that there is another way that cooperation can become entrenched in a society: through the "you scratch my back and I'll scratch yours" principle or, to put it more technically, through reciprocal altruism. Reciprocal altruism can evolve in societies where the costs of helping one another are less than the benefits received and where individuals can keep track of who has done what to whom. Switch on the TV and you won't have to watch for long before a sit-com or soap opera picks up the theme of who has done what to whom. We humans are obsessed with reciprocity. Gossip (which is all most TV programs amount to) is a time-consuming process of keeping tabs on each other. Could any other animal be keeping track of who does what to whom?

One of the few species in which reciprocal altruism has been clearly identified is the vampire bat. Vampires live together in large roosts in hollow trees (a single male at the top with his harem beneath him). In these groups, foraging mother bats commonly regurgitate blood for their young. Though touching, this display of maternal affection does not get biologists very excited—after all, the young share half their genes with their mothers, and most mammal species show some maternal care for offspring. This is simply genetic nepotism.

Much more exciting is the discovery made by Gerald Wilkinson, now at the University of Maryland, that adult female vampires also spit up blood for other mature females to whom they are not related. Vampires that have failed to find a blood meal make a special begging sound to their fellows in the roost. By grooming each other, replete bats can feel if the beggar really has an empty stomach and, if so, regurgitate a donation of blood for her. This generosity costs the donor little (a replete bat has more than enough to make it to the next night), and it saves the life of the recipient. Moreover, Wilkinson demonstrated that bats are more likely to donate to a bat that has helped her out in the past: they remember who has helped them previously and repay those favors. It is this friendly reciprocity that enables vampires to survive up to twenty years in the wild.

Amid all the strange habits of bats, here at last we see a hint of something we can identify with. If a colleague has come to work without his wallet, I might lend him money for lunch. I don't even insist on feeling his stomach first to check if he really needs another meal. But I'm unlikely to forget my generosity: reciprocity is almost always implied in human kindnesses. Although we might draw the line at regurgitating a meal we'd already eaten, vampire bats, in their charity, are more like us than we might ever have guessed. Bats are not always alien; they can be strangely familiar too.

WHAT IS IT LIKE TO BE A PHILOSOPHER?

I have shared dinner with philosophers, and drunk their wine. Though I respect them greatly (some of them, at least), I do not understand how philosophers think. We may never know what it is like to be a bat. Indeed, it may well not be like anything to be a bat. But to me the struggle to understand objectively what bats are, using the most reliable methods we have available, the highest technology and smartest training methods, is a noble quest. And its result, an ever clearer picture of the world of the bat, is a thing of beauty, the equal of a work of art. And, what is more, this knowledge may help us to aid bats—to undo some of the damage we have done them over the years.

Bats have had a hard time. Insecticides accumulate in the bodies of insect-eating bats. Fruit farmers in Australia fry flying foxes on electric wires to stop them getting at fruit crops. With the destruction of their forest habitat, bats in Europe have taken to sharing human structures, such as the lofts of houses, disused churches, and farm outbuildings, and proximity has not endeared them. Another British novelist, the famously fusty Auberon Waugh, wrote to a national newspaper to decry attempts to preserve endangered species of bat: "I do not suppose that there are more than a couple of hundred people could give a hoot if every bat in the kingdom dropped down dead. I, for one, would rejoice. . . . Like horseflies, they have absolutely nothing to recommend them. They are dirty, smelly and frightening." Ancient superstitions about evil bats are a long time dying. In 1922 a Sussex farmer was caught nailing a bat upside down above his barn door to keep harm away from the farm. As recently as 1988, police in North Wales investigated a complaint along the same lines.

Perhaps paradoxically, bats seem to get a better press in parts of the world where they are eaten than in places where they are not. Although Mosaic law forbids the eating of bats by Jews, people in many parts of the world, especially Africa and the Far East, regularly eat some of the larger fruit-eating species. An Australian settler, copying her aboriginal neighbors, tried cooking a flying fox and reported, that "the flesh when they are in season very much resembles sucking pig." In China bats are held to bring good luck: the word for bat, *fu*, is also the character for happiness, and bats have been credited with supernatural powers and even given the status of minor gods. A talisman showing five bats represents the five greatest joys to mankind (which, in case you didn't know, are health, happiness, longevity, prosperity, and contentment).

Islam too is also more tolerant of bats than Judaism and Christianity are. There is a Muslim belief about the creation of bats by Christ. During Ramadan, when all must fast between sunrise and sunset, Jesus went up into the hills around Jerusalem. Here the mountains shut out the western sky and made it impossible to be sure when the sun had set and therefore when one might eat. Jesus took some clay, formed it into a winged creature, prayed, breathed on it, and immediately it flew away into a cave in the mountains. Every night around sunset this new winged beast would flutter around Jesus and apprise him of the hour.

But even in the West a countermovement is developing to our traditional fear and hatred of bats. People are mobilizing to protect the mammals of the air. In Britain there are Bat Groups in every region of the country, and in the United States and elsewhere there is Bat Conservation International, headquartered in Austin, Texas. We may never know what it is like to be a bat, but we can still stand in awe of them. So long as we ensure that there are bats left to wonder at.

Bats are harmless and wonderful. The Western superstitions and fears about them are unfounded. Even the vampire only takes half an ounce of blood at each feeding. Other bats rid us of insects and pollinate commercial crops. The Eastern philosophies that take bats as symbols of good luck and prosperity really are closer to the mark than the odious Western tradition of viewing bats as evil.

The things bats do are wonderful adaptations to the niches they inhabit. Unremitting natural selection can shape mechanisms, like echolocation, so complex that it takes centuries of human ingenuity to comprehend or emulate them. Bats seem strange to us but no more so than we would seem to them, if they had anyway to know us. Let's not worry about what they might be conscious of: let's just enjoy our opportunity to be conscious of them.

FURTHER READING

Bats, by M. Brock Fenton (Facts on File, 1992). A wonderfully illustrated, clearly written, and marvelously informative volume about all bat species.

The Biology of Bats, by Gerhard Neuweiler (Oxford University Press, 2000). An excellent, up-to-date, advanced introduction to the scientific facts about bats.

Bats, by Glover Morrill Allen (Harvard University Press, 1940). The science in this book is way out of date, but the folklore material is timeless.

Bat Conservation International. <www.batcon.org>. This web site is a terrific compendium of information about many species of bat, as well as a clearinghouse for information about bat conservation efforts around the world.

5

Talk to Me

Language is the crux of the matter. Beings like us talk to each other. The others don't. For centuries people have recognized a sharp distinction between themselves and all other species when it comes to language. We do it; they don't.

In the last few decades, however, bolder claims have been made for the language abilities of some of our closest relatives, such as chimpanzees, bonobos, and gorillas, as well as for such diverse beasts as dolphins, parrots, and honeybees. In this chapter I want to take a careful look at these claims. Has modern science uncovered what the ancients failed to notice, a capacity for language hidden in other species?

We have already seen, in chapter 2, that honeybees can communicate with their hivemates about food sources. So we know that some form of communication is possible between nonhumans. But is this language?

Language is a special case of communication. With our languages we can communicate a range of ideas not even limited by our vocabularies—large as they may be. Educated people have vocabularies in the many tens of thousands of words. But with the addition of grammar, a system of rules for combining words, our communicative potential becomes truly unlimited.

Those of us who make our living from words can be accused of overemphasizing the importance of language. As I was doing my research for this chapter, three workmen dug a trench across the road outside my window. One of them drove a digger while the other stood in the ditch and shoveled out any dirt left by the digger. From time to time the third would wander over to see how things were going. It's hardly an intellectual pursuit. This is back-breaking manual labor. But every now and then, about once every ten to fifteen minutes, they stopped to discuss the work. I couldn't make out what they were saying, but it was clear from the way they pointed things out to each other that they were sharing information about the ground they were digging through, about the cables to watch out for, and other obstacles under the street. This ability to communicate about objects and their interrelationships enables them to work in far closer collaboration than if they had no common language. This ability to locate objects and their relationship to actors ("You dig here, I'll dig there") stems directly from the syntax of language. It is easy to see how early hunters (or hunters today) could gain an advantage from this ability.

Notice how the dance communication system of the honeybee does not have this open-ended quality. Bees cannot recombine the elements of their dances to communicate about a beetle infestation. They cannot dance backward to indicate a place best avoided. Though they have a way of communicating, it is a constrained one.

Steven Pinker is professor of psychology at MIT, and one of the world's leading authorities on the psychology of language. He is also blessed with great skill in conveying his knowledge on the subject to a broad audience. To take just one key point from Pinker's *The Language Instinct*: many of the sentences you say have never been uttered before in the history of humankind. That is how flexible our language is. Some tens of thousands of words combined with the rules of grammar (each language has its own rules, but all grammars have some things in common), and the result is effectively infinite flexibility. If you don't believe me (or

Pinker), try this. Open up an Internet search engine and search for the first five words of any sentence in this book. I make no claims that my prose is particularly novel, and yet if you search that most massive library of texts we call the World Wide Web (presently around 3 billion pages), you will find that even my text is original enough that none of my sentences are already in there. (If you try this, you need to put the words in inverted commas or quotation marks so that the search engine searches for those words *in that order* with no other word intervening. Otherwise search engines just search for those words anywhere on the page.) I chose the phrase "Notice how the dance communication" from the beginning of the paragraph before last. It sounds pedestrian enough—surely others have used it. "Notice how" and "Notice how the" both came up about 122,000 times. "Notice how the dance" is on one web page. But "Notice how the dance communication," bland as it sounds, has never been posted on the Internet. That doesn't prove, of course, that it has never been said before, but remember that it includes just the first five words of my sentence. By the time that sentence ends, eleven words later, it may well be truly unique. If you want to check that my search strategy really would find those words if they were out there, try a familiar phrase (I chose "the cat sat on the mat"): these usually come up several thousand times.

There are two ways that we can look for language in other species. The first is to take a beast into our homes and laboratories and try and teach it a communicative system of our devising. Spoken English, American sign language for the deaf, and other systems of button pressing or hand waving designed specially for beast to man communication have all been tried. The alternative method is for the human to enter the home of the beast. Some researchers have actually set up camp among free-living animals in Africa and elsewhere; others are content to study animals in the laboratory. In either case no training is attempted: the animal's spontaneous communicative skills are the object of study.

I want to start by considering the various attempts that have been made over the last century and a half to train animals to use human language. The critical question to bear in mind is, Has any animal succeeded in learning an open-ended language system like our own, or have other species only mastered communication in a more closed manner, like the honeybee's communicative dance?

APING LANGUAGE

With the opening of the London Zoological Gardens in Regent's Park in 1828, Victorian England got a closer look at exotic creatures than ever before. The apes, with their obvious similarities to human beings, impressed many observers. Queen Victoria exclaimed, "The Ourang-Outang is too wonderful. . . . He is frightfully and painfully and disagreeably human." Charles Darwin was also fascinated by the orang's similarity to a human being: "Let man visit Ourang-outang in domestication, hear expressive whine, see its intelligence when spoken [to]; as if it understood every word said—see its affection—to those it knew—see its passion and rage, sulkiness, and very actions of despair; and then let him dare to boast of his proud pre-eminence."

Indeed, so taken was Darwin with the apes that he prevailed upon his close follower George Romanes to try the experiment of adopting a chimpanzee to see if it could be taught to speak. Romanes was not notably successful, but others, not discouraged, made the same attempt. In the years leading up to the First World War, an American investigator, William Furness, tried training chimpanzees and orangutans but was never able to get more than a single "Mama" out of one chimpanzee and a couple of recognizable words from the orangs. In the 1920s another American, the influential pioneer animal psychologist Robert Yerkes, tried to train two chimpanzees, Chim and Panzee, to speak, but the same lack of success greeted his efforts. Yerkes noted that "perhaps they can be taught to use their fingers, somewhat as does the deaf and dumb person, and thus helped to acquire a simple non-vocal "'sign language.'" This sensible suggestion went unnoticed for another fifty years.

Undeterred, Winthrop and Luella Kellogg, then at Indiana University, adopted Gua, a one-year-old female chimpanzee. They

"Ready for bed": Winthrop Kellogg's son Donald at 18½ months holding hands with Gua (age 16 months). (Archives of the History of American Psychology, Winthrop R. Kellogg Collection. Courtesy of Shirley Kellogg Ingalls and Patricia D. Kellogg)

moved down to a large primate research center that Yerkes had opened at Orange Park, just outside Jacksonville in the northeast of Florida. Winthrop and Luella raised Gua alongside their son of the same age for about nine months in 1931 and 1932. The Kelloggs hoped that by raising Gua in their home so closely with their son, who was just at the age when human children learn to talk, they might succeed in instilling language in the chimp. Winthrop was particularly keen to get Gua to say "Papa." He laid the chimp on his lap and "slowly and distinctly uttered the syllables "'pa-pa.'" When this didn't evoke any response, Winthrop tried manipulating Gua's lips as he said the word. The only outcome of this training was that, after a month or two of manipulation, Gua would stick her own index finger between her lips whenever Kellogg said "Papa." After a fruitless nine months, the Kelloggs packed up and went home to Indiana.

The most successful of the early ape language studies was that carried out by Cathy and Keith Hayes in the late 1940s. Like the Kelloggs before them, the Hayes found their chimpanzee, Viki, at the Orange Park primate center in northern Florida. Cathy Hayes's account of their adoption of Viki stands out as an exquisite memoir, not just of a very challenging experiment in cross-species adoption, but also of life in middle-class America in the late 1940s. Particularly striking is the combination of her very real affection for her ape infant with her clear-headed realism about how little they were able to achieve in terms of teaching the ape to communicate. Once again the ape was taken on its foster-parent's knee and encouraged to say "mama" and "papa."

Cathy Hayes noted that Viki did not babble like a human infant. Around three months of age, human babies start to make a whole range of sounds—sounds that are part of the language they hear around them, but also other sounds found in other tongues. As the infant grows, he gradually drops the sounds not used in the language of the people he has been born into and concentrates on practicing just those sounds his own linguistic community actually uses. But Viki did not babble. She made a few sounds, particularly in response to emotional states like fear or when she was demanding food, but she did not play with her voice as a human child does. Cathy Hayes reported: "There were a few encouraging little flare-ups, like the day she went Hawaiian with remarks like 'ah ha wha he' and 'ah wha he o.' But these exceptions only made the next day's silence more discouraging."

The only way the Hayes could find to get Viki to speak was to push her lips together while she was emitting her "asking sound," which was "ahhh." By pressing and releasing Viki's lips in this way, the Hayes were able to trick the ape into saying "Ma, ma." Impatient for the rewards that came her way for producing the desired sound, Viki would press her own fingers to her lips while saying "ahhh." Two weeks after the first training, Viki was able to move her lips without the assistance of her own or anyone else's fingers and said her first unaided "mama." Though they went on in similar ways to train Viki to say "papa" and "cup," the Hayes were never the least deluded about the limitations of what they had achieved. This clear-headedness did not prevent them being amused at the delusions of others:

> A policeman came down the street, placidly strolling his beat. He spied Viki. He came over and talked to me, and he and Viki exchanged broad grins. Suddenly, for a reason I will never know, she said very loudly and distinctly, "Cup!"
>
> The policeman straightened up. "She said, 'Cop'!" he exclaimed.
>
> "She did?" I replied, amazed at his interpretation.
>
> "Well, didn't you hear her? She said, 'Cop'! I'll be flat-footed!" And then eagerly, "Can she say anything else?"
>
> Seeing that his standards were so lax, and his pleasure so genuine, I could not resist the temptation to shrug and say, "Oh, she says lots of things."

Though Cathy Hayes finished her book, published in 1951, with high hopes of what she and her husband might achieve with Viki, (the final chapter is titled, "It's Only the Beginning"), only a few years after publication of the book the Hayes were divorced, and, tragically, Viki died of encephalitis in 1954.

By that time, a consensus was developing that apes were not capable of language. The ancients were right: language was a uniquely human gift after all.

JUST SHOW ME A SIGN

But Allen and Beatrice Gardner of the University of Oklahoma had an original plan. Instead of trying to get their adopted chimpanzee,

Washoe, to speak in vocal words, they trained her in sign language, specifically, Ameslan, American Sign Language for the deaf. Astute observers, from the great seventeenth-century English diarist Samuel Pepys on, had noticed chimpanzees' enthusiasm for what look to us like hand gestures. Pepys and, three centuries later, Yerkes suggested that it might be possible to communicate with apes using signs, but nobody had actually tried it before. Washoe lived in a trailer in the Gardners' backyard and was surrounded at all times by human trainers who communicated with her and with each other in Ameslan. Though Washoe was not inclined to spontaneously imitate her human companions' signs, she readily learned to sign when her hands were molded into the desired positions.

By the time the original Project Washoe ended in 1970, Washoe could, after four years' training, reliably produce 132 signs in appropriate contexts. The Gardners' achievement was a colossal breakthrough in ape language studies. Where attempts over the previous century had never succeeded in getting more than three words from a chimpanzee, Allen and Beatrice Gardner had pushed the limit out to over one hundred words. With this larger vocabulary, the Gardners tried to test Washoe's comprehension of structural aspects of language, such as syntax and grammar. They also noted Washoe's originality with language. In an often repeated anecdote, Washoe, taken out on a boat on a lake, saw a swan for the first time and signed, "water bird."

Suddenly what had been the standard view was overturned. Prior to the Gardners' research, the prevailing position was that chimps were incapable of learning human language because they lacked the specialized brain structures that underpin its comprehension and production. With the publication of Washoe's feats, the new received wisdom became that chimpanzees only lacked the ability to *speak*: they lacked the human specializations that control breathing and activate the vocal chords. But a latent talent for language could nevertheless be uncovered; it was just a question of finding the correct form of expression—and that was sign language.

The Gardners' success with Washoe inspired several imitators. The most significant of these was Herbert Terrace of Columbia

University in New York, with his chimp, Nim Chimpsky. Nim was named in joking honor of the great linguist Noam Chomsky.

During the 1960s and '70s, a furious debate raged between nativists and environmentalists over the development of human language: the former supported nature and genetic contributions as being critical in the development of language, while the latter argued for the effects of nurture and the environment. Foremost on the side of nature was (and still is) Noam Chomsky at the Massachusetts Institute of Technology. Chomsky argued that the efforts parents make in teaching language to their children were so obviously insufficient to the task that an ability to use language must be innate in the human infant; otherwise we couldn't possibly pick up our native tongue. From this point of view, it was highly improbable that anything like human language would be demonstrated in another species. The other extreme was represented by B. F. (Burrhus Frederick—Fred, to his friends) Skinner, the father of modern radical behaviorism, down the road from Chomsky at Harvard University. Skinner argued that language, like any other behavior, was learned by the individual through interaction with the environment. Skinner published a book in 1957 in which he talked of language "acts" as if they were like any other kind of behavior, as if they could be compared to the pecking of a pigeon or a the running of a rat in a maze. Chomsky retaliated with a famously critical review of this book in 1959.

Terrace had been a student of Skinner's at Harvard. He named Nim in honor of Chomsky because it would be a wonderful joke if the chimp that refuted Chomsky's contention that language was unique to humans bore the great man's name.

The Gardners' success in training Washoe supported Skinner and his camp in their debate with Chomsky. If an ape could be taught sign language, then, by implication, human language might owe more to the environment than to hereditary factors. The chimp was still descended from a long line of chimps; only his environment had been changed.

Over a three-and-a-half-year period, Nim was taught to use some 125 Ameslan signs. At the time, Nim seemed a pretty close copy of Washoe: he was brought up with intense human contact, just as she was, and mastered 125 signs in three years as compared to her 132 in four, a pretty similar result. At first the two cases were treated as equivalent. But after the initial training period, the stories of these two apes started to diverge rapidly.

There is no inexpensive way to keep a chimp in Manhattan, so when Terrace's funding for Project Nim ran out, he had no choice but to give Nim away. Freed of the burden of caring for the chimp, Terrace took the opportunity to carefully analyze the many hours of videotape that he had collected of Nim's signing activities. What Terrace uncovered caused a great storm. By the time his book, *Nim*, appeared in 1979, Terrace had concluded that what Nim (and by implication the other apes in language projects) was doing had little to do with language as we normally understand it. Instead, said Terrace, the chimps had achieved a simpler form of learning: that making certain signs led to certain consequences. The chimps had learned to produce certain arm and hand movements to demand things they wanted: "I do this; I get that." This kind of learning about cues and consequences, as we saw in chapter 3, is a part of the filling of the similarity sandwich—a capacity that most species share.

The first thing that Terrace noted was that the vocabulary of these trained apes, although far larger (up to 250 words) than had ever been achieved in the attempts to get apes to speak vocally, was, in fact, a very modest accomplishment compared to that of human infants learning their first language. Healthy two-year-old children learn nearly ten new words every day—about one every waking hour. Washoe, Nim, and the other apes in language training never came close to this "spurt" of language learning in the human child. The largest vocabulary claimed for any of the language-trained apes is that of the gorilla Koko, who is said to know over one thousand words. Francine Patterson has trained Koko for over twenty-five years, using the methods pioneered by the Gardners. How does one thousand words compare to a human vocabulary? Though it is always fashionable to bemoan the limited vocabulary of contemporary youth, the average U.S. high school graduate knows around forty thousand words. A decent dictionary counts its entries in the hundred of thousands.

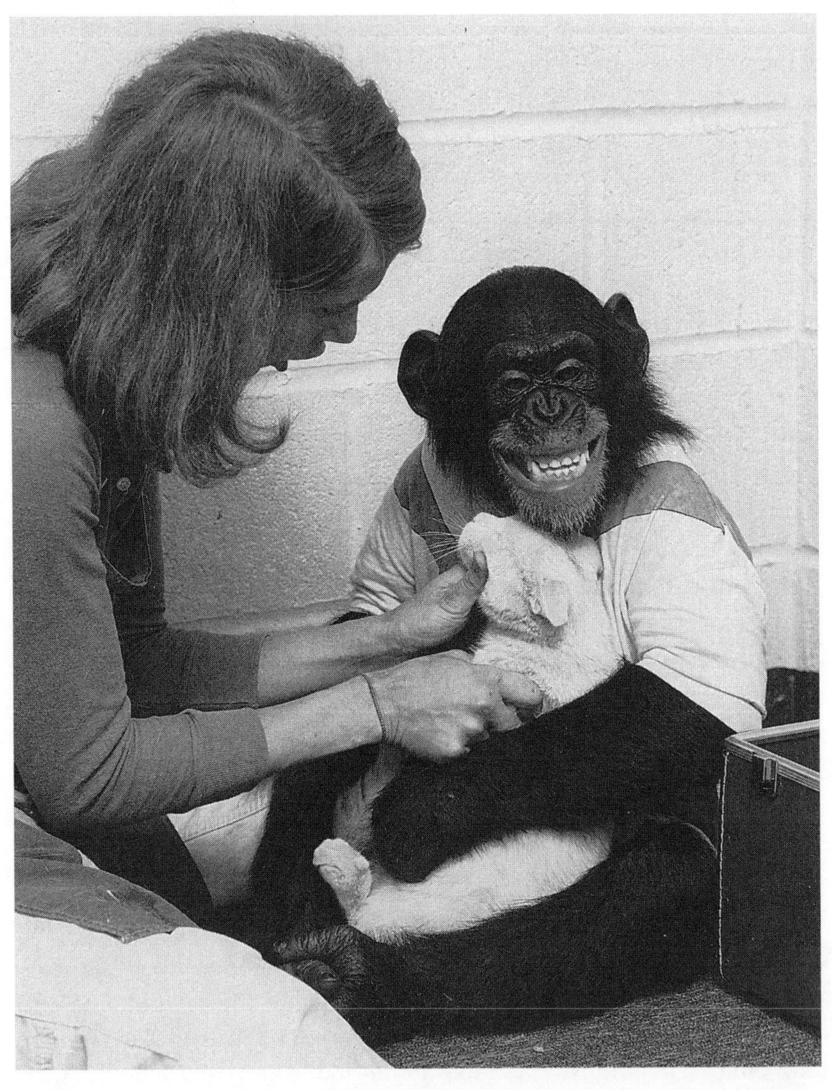

Nim cuddling a cat in his "sterile" (Steven Wise's term) testing room at Columbia University. (H. Terrace, Columbia University)

Terrace did not just point out Nim's limited vocabulary and failure to show a "spurt" of vocabulary development. He also noticed that Nim's utterances did not increase in length.

Human children start by just naming objects; then they soon come to add adjectives to their nouns, and by the third year of life are stringing together little sentences. This never happened to Nim. The average length of his utterances remained stuck at only a little over one word throughout his entire training period. And when he did occasionally say something a bit lengthier, it showed no grasp of the demands of grammatical structure.

Now grammar is a funny thing. Many of us, though we may consider ourselves educated people and able to string together a sentence or two when required, would not claim any knowledge of grammar. And when we do hear a grammatical concept articulated, it often sounds fairly arbitrary and perverse. As Winston Churchill said to the civil servant who attempted to reword the great man's writing to avoid terminating a sentence with a preposition: "This is the kind of bloody nonsense up with which I will not put!"

But we do all know grammar and use it every day. It is grammar that makes the difference between "Man bites dog" and "Dog bites man"—a distinction few three-year-olds and no five-year-olds would have trouble understanding. In English, word order carries most of the weight of disambiguating who did what to whom. In other languages, other methods, such as changing word endings, may do the same work. However it is done, it is crucial to language that actors be differentiated from things acted upon. Have any of the language-trained apes shown any sensitivity to grammar?

Nim's longest utterance is not hard to understand: "Give orange me give eat orange me eat orange give me eat orange give me you"—but neither is it in any way grammatical. In fact it is typical of Nim's longer statements in that it is lengthened simply by meaningless repetition. It is also typical of Nim's utterances in that it is a demand for something. Nim did not go around naming the objects in his world the way children do, and neither did any of the other language-trained apes. Nim used words to express wants, but that is as far as he was willing to go in the art of communication.

From close inspection of the videotapes, Terrace noted that when asked a question, Nim often exploited the signs used in the question to formulate his answer. Furthermore, analysis of the tapes uncovered an unconscious tendency on the part of the human signer, while waiting for his response, to offer Nim the signs that he would need for it.

Rare indeed is the scientist who publishes such strong criticisms of his own research. Terrace's critique was an enormous shock to the ape language community. The first response of the other ape language researchers was to claim that Terrace's problems couldn't possibly apply to their own research. Terrace was criticized for having permitted too many people to work with Nim. By having so many caretakers, they claimed, Nim had failed to form the kinds of strong bonds to individual people that were needed if true language competence was going to show itself. Terrace included in his book photographs of the special room where the critical tests on Nim's language competence were carried out. Some of his critics argued that Nim had been brought up in a severely impoverished, prisonlike environment—so no wonder he had failed to develop language.

These criticisms were highly selective. All the ape-language projects have had a large personnel. The need for round-the-clock animal care, seven days a week for several years, makes it all but impossible to avoid calling on help from a long list of assistants. And Terrace's critics ignored the fact that the austere room pictured in the book was not where Nim had lived but only a special language testing room designed to minimize distractions for the lively little ape. In the early days of Project Nim, the chimp had lived with a family in a Manhattan apartment. When he became too rambunctious for condo life, he was moved to a magnificent mansion. If we were going to rank the luxuriousness of the living quarters occupied by apes in language training, Nim would win any competition, hands down.

Perhaps the strangest criticism fired at Terrace was that he was a behaviorist and consequently only out to prove that the signing apes' apparent language was nothing but a set of behaviors trained to obtain reinforcement. We have already seen there could be little doubt about Terrace's behaviorist credentials. But far from explaining why Terrace would become such a trenchant critic of the ape language studies, it was precisely his leaning toward the behaviorist, or environmentalist, side of the debate between Skinner and Chomsky about the origins of language that had inspired him to take up the challenge of training Nim in the first place. In this he was little different from the others involved in the ape language projects of the 1960s and 1970s. Consequently, the attacks on Terrace as a behaviorist who would be temperamentally disinclined to accept evidence of language in apes have the intellectual climate of the time completely round the wrong way. Considering his background as Skinner's student, we would expect him to champion the ability of apes to learn language. His about-face is therefore all the more striking.

One of the strongest critics of Terrace's work with Nim is animal rights advocate Steven Wise. In his book *Rattling the Cage*, Wise accuses Terrace of not caring enough about Nim for language to develop: "what he actually proved was that with enough chaos and with poor enough teaching, a chimpanzee can nearly be prevented from learning ASL [American Sign Language]." Terrace, according to Wise, loved Nim neither wisely nor well enough. Trying to build up a picture of Terrace's unwholesomeness, Wise also describe him—falsely and irrelevantly—as a lifelong bachelor.

KANZI: "THE EINSTEIN AND SHAKESPEARE OF THE CHIMPANZEE WORLD"

Thirty years on from the first ape language projects, there are few animal psychologists left who would defend the idea that those original studies demonstrated anything like human language in their charges. David Premack (who trained a chimp called Sarah at the University of California, Santa Barbara) and Terrace do not claim that their apes learned language. Francine Patterson, the trainer of Koko the gorilla, and Roger Fouts, who now has custody of Washoe, have all but disappeared from the scientific literature (though they still appear with enthusiasm in the popular media). There is only one ape for whom serious claims of linguistic competence are still made in scientific journals.

Around the same time that Terrace was working with Nim, Duane Rumbaugh of the Yerkes regional primate center developed a computerized communication system where the ape had to press keys on a specially designed keyboard to indicate its responses. These keys were covered with symbols known as lexigrams. Lexigrams have symbolic value in the artificial language system that Rumbaugh designed and called "Yerkish" in honor of Robert Yerkes. Just as the word "apple" has nothing that looks, sounds, or tastes like an apple about it, so the Yerkish lexigram for apple (a blue triangle) is completely arbitrary, with no "apple like" qualities.

Sue Savage-Rumbaugh—Duane Rumbaugh's sometime student, sometime wife and collaborator—decided to try training a new species of ape using the lexigrams. Bonobos (also known as pygmy chimpanzees) are so closely related to chimps that until recently they were thought of as a subspecies of chimpanzee. Though bonobos look rather like chimpanzees (they are slightly smaller and more likely to stand on their hind legs like humans), they also differ from chimps in a number of interesting ways. For one thing, bonobos are much less violent. Their societies are dominated by the females, and sexual competition between males seems to have been written out of the equation. Sexual gratification is shared freely, and most aggressive encounters end not with bloodshed but with ritual copulation or mutual masturbation—including between animals of the same sex. Though bonobos may not quite be the loving apes that some web sites would have us believe, a friend of mine who has worked with Kanzi and has bite marks on his shoulder as a souvenir of their acquaintance, nonetheless told me that, "by comparison to chimps, rhesus, or humans with pointed sticks, bonobos really are gentle."

Savage-Rumbaugh started out trying to train an adult female bonobo, Matata, to communicate using the lexigram keyboard, but to no avail. Matata was a hopeless student. All the time Matata was in training, she had her adopted son Kanzi hanging on to her body. After two years of failure with Matata, Savage-Rumbaugh decided to remove Matata from the scene and test Kanzi instead. According to Savage-Rumbaugh, in the absence of any explicit training, Kanzi demonstrated a spontaneous understanding of the use of the Yerkish lexigrams. He correctly used the keyboard to indicate items he wanted and to name things he was shown.

It is worth pausing for a moment to contemplate what is being claimed for Kanzi. According to Savage-Rumbaugh, Kanzi, the adopted child of a mother who never learned to communicate, acquired a language without any direct training. What human child learns language without any communication from his parents or any other opportunity to observe language in action? Surely we wouldn't expect an ape to learn language under conditions where a human infant would not. Why would Savage-Rumbaugh want to make such an extreme claim for the bonobo in her charge?

The motivation for this claim surely came from the changing intellectual climate. By the mid-1980s few wanted to defend Skinner's environmental account of language development against Chomsky's view that language ability is innate. Thus, to claim that Kanzi's communications were "real" language, it seemed important to be able to argue that environmental contributions to Kanzi's abilities were minimal. The fact that Savage-Rumbaugh has created an account of a bonobo that learned language under conditions where no human infant would succeed has not gone unremarked. Small wonder that Robin Dunbar from the University of Liverpool calls Kanzi "the Einstein and Shakespeare of the chimpanzee world rolled into one."

It's hard not to get frustrated at these debates about whether the language-trained apes really understand language. Maybe we can't all have a one-on-one opportunity to converse with a gorilla or bonobo, but just show us the transcript; let us decide for ourselves whether we accept this as language. Well, the ape language supporters guard the transcripts of their apes' utterances better than the biological weapons labs guard their anthrax—you just can't get one. When Terrace analyzed Washoe's utterances from videotape, the Gardners threatened to sue for breach of copyright.

The only specimens of ape language in the public domain are transcripts of a TV documentary and an ape on the Web answering questions put to it by email. These are highly unsatisfactory, because they are edited products of the entertainment industry: they were produced to support claims that these animals have mastered language and to provide good entertainment. No independent scrutiny of their conditions of production was possible.

And yet even these sources are enough to prove my point: none of these animals have mastered language.

Here is a transcript of an interaction between Kanzi and his trainers, recorded by a Japanese TV crew. Kanzi communicates by selecting keys from his keyboard, which contains a "dazzling" (Steven Wise's term) 250 keys. His trainers communicate to him in spoken English. This was recorded in 1993, when Kanzi was thirteen years old and had been in language training for eleven years:

> HUMAN: Kanzi, this is Janine. Would you like any food? Tell me what food you'd like.
>
> KANZI: Food surprise.
>
> HUMAN: Some food surprise?
>
> KANZI: Food surprise.
>
> HUMAN: Kanzi, would you like a juice, or some M&Ms, or some sugar cane?
>
> KANZI: M&Ms.
>
> HUMAN: You like M&Ms? Okay. Kanzi, is there any other food you'd like me to bring in the backpack?
>
> KANZI: Ball.
>
> HUMAN: A ball? Okay.

Look at this exchange carefully. First consider the human's speech. I don't think we'd consider her language complex; it's the kind of thing you might say to a young child. And yet she uses an average of ten words per utterance, while Kanzi's responses average just one and a half words. (One and a half words is pretty standard for ape speech.) The human identifies herself; she asks questions; she comments on the ape's responses. Kanzi just names things in single words, except for two uses of the extra word "surprise." This addition might indicate that Kanzi wants to be surprised by whatever food he gets (though if that's the case, why does he ask for M&Ms when the same question is put to him a different way?), or it might be completely meaningless. Kanzi doesn't offer any further explanation of himself.

Although Janine responds to the ape with some questioning repetition that attempts to draw out a clearer answer, she is not simply echoing Kanzi's language. Kanzi, on the other hand, in those cases where his answers make the most sense, is simply drawing his response from something that Janine said, for example, "food"

Kanzi out in the woods with his keyboard and Sue Savage-Rumbaugh. (Nicholas Nichols, National Geographic Image Collection)

or "M&Ms." When Kanzi adds vocabulary to the conversation, he is always doing so at the expense of comprehensibility, as when he uses "surprise" and "Ball." In the last interchange, "Ball" can be understood to mean "I don't want any more food in the bag, but I'd like a ball instead," but again Kanzi doesn't take the opportunity to explain this for us. The simplest explanation of what he has said is that he is ignoring Janine's question and just demanding something he wants.

Here is everything else Kanzi says in this documentary:

KANZI: Want milk. Milk.

HUMAN: You want some milk? I know, you always want some milk when you're planning to be good.

KANZI: Key. Matada. Good.

HUMAN: Oh, you want the key to Matada, and you're going to be good. Well, I'm glad to hear that. I'm glad to hear that.

Note how abrupt, demanding, and/or meaningless Kanzi's utterances are on their own and how his trainer struggles to add comprehensibility to them. "Key. Matada. Good" means "You want the key to Matada, and you're going to be good." Note too how the transcriber has added periods after almost every word that Kanzi utters. These words just refuse to string together into sentences.

Nevertheless, Savage-Rumbaugh also claims that Kanzi is the first ape to understand grammar. She and her team tested Kanzi's understanding of syntax with 660 different sentences, such as "Would you please carry the straw." Of these 660, Kanzi responded correctly to 72 percent: substantial evidence, claim Savage-Rumbaugh and her supporters, that he understood the grammatical implications of these sentences and deduced who was to do what to whom. Unfortunately, this test is much weaker than it at first appears.

Most of the sentences on which Kanzi was tested were not at all ambiguous as to what should be done to whom. In carrying out the instruction "Would you please carry the straw," Kanzi may simply have understood "you," "carry," and "straw" and formed the only plausible conclusion from those three tokens. Kanzi can carry a straw, whereas a straw will not carry Kanzi. The most direct test for syntax comprehension would be to try both versions of a reversible statement, for example, "Put the ball on the hat" and "Put the hat on the ball." This is the best test of whether an individual can understand grammar: give him statements that demand grammatical competence for their comprehension. Unfortunately, out of the 660 sentences used with Kanzi, there were only twenty-one pairs of reversible statements. Savage-Rumbaugh and colleagues report Kanzi as responding correctly on twelve of these twenty-one pairs—a modest 57 percent. However, these twenty-one pairs include three that were presented during an early phase of testing in which no attempt was made to prevent the ape from picking up on bodily cues from the experimenter, such as eye movement. We know that apes attend very closely to signs that a trainer may unintentionally give by slight eye or body movement. Without these three pairs, Kanzi's score drops to nine out of eighteen, or 50 percent.

But the system by which Savage-Rumbaugh and her colleagues

scored Kanzi's reactions to their instructions is extremely generous. They commanded Kanzi, "Pour the juice in the egg." Kanzi picked up the bowl with the egg in it, smelled it, and shook it. Only after they repeated the command three times with various modifications did Kanzi do what they asked him to. They nonetheless score this as a correct response. Similarly, although Kanzi's first response to the demand "Pour some water on the raisins" was to hold a jug of water over a lettuce, this was coded as correct. Kanzi's first reaction to the request to pour milk into water was to stick a tomato in the water. When asked to chase Liz, he remained seated; when asked again, he touched Liz's leg, and she chased him. His response in all of these instances were scored correct. When Kanzi is given the two commands "Make the [toy] doggie bite the [toy] snake" and "Make the snake bite the doggie," in both cases the snake ends up in the dog's mouth, but both responses are scored as correct. Rescored to exclude these "false positives," Kanzi achieves correct responses to only five pairs out of eighteen: less than 30 percent.

This might just seem like nitpicking. Why should we be concerned whether Kanzi makes the right response straightaway so long as he gets around to doing something like what he was asked to do sooner or later? But this test is the only evidence we have for whether or not any nonhuman comprehends grammatical utterances. And grammar is what makes the difference between being able to express a number of ideas equal to the number of words you know and being able to express any idea whatsoever. We already know that Kanzi has in his vocabulary the words used in this test. We are not interested in demonstrating that if we say "Make the snake bite the doggie," Kanzi will do something involving the dog, the snake, and biting. It is absolutely central to this test that he make the snake bite the dog, not make the dog bite the snake, or bite one or both of these toys himself, or bite something else. Only one response can show grammatical comprehension. That is why a strict criterion for scoring success is absolutely essential. And on a strict criterion, Kanzi just doesn't get grammar.

Nor is Kanzi's fluency with his "language" any better than that of his predecessors. Just as with the previous "language-trained" apes, the vast majority of Kanzi's utterances (94 percent) are just a single sign (his utterances average 1.15 signs), and his vocabulary remains in the low hundreds.

In recent years, Kanzi's lexigram keyboard has been connected to a voice synthesizer chip. Consequently, any English-speaking human being can put a question to Kanzi and understand his response. Several journalists have published interviews with Kanzi, and those that listen to what Kanzi says, rather than Savage-Rumbaugh's interpretations, notice the ape's tendency to monosyllabic conversation focused on food treats.

Supporters of the ape language projects like to compare their charges' use of words to the behavior of children who are only just starting to use language. But do they ever go beyond this initial phase? Human children rapidly move on to more sophisticated utterances. Consider this utterance of a Greenlandic two-year-old: "tattuus-sinaa-nngil-angut" (we cannot be so crowded together in it). Though clearly precocious, this youngster is only saying sooner what all Greenlandic people are capable of expressing by the time they are three or four years old. By age five, normally developing children have the complete apparatus of their native language at their fingertips—minus a few finer points of grammar and a full-size vocabulary.

For all the excitement and all the TV documentaries, the so-called "language-trained" apes have not learned language. As Steven Pinker says in *The Language Instinct*, they "just don't 'get it.'" They sign or press buttons because doing so gets them what they want. They can be drilled to string a couple of signs together but usually can't be bothered. Although some of them have been in training for decades, there is nothing to suggest that any of them ever comprehend grammar. Grammar is the crucial lubricant that opens language up from being limited by our vocabulary to being completely infinite in its expressive possibilities.

TALKING TO THEMSELVES

Why should evolution have led to nonhumans that can communicate with humans? When researchers tried to teach human language to apes, they were working on the assumption that genetic relatedness would dictate how likely two species would be to communicate with each other. Humans and chimps are closely related; ergo, they are likely to understand each other. But this is a fallacious argument. Aside from our domestic companions, such as cats and dogs, there has been no evolutionary selection pressure on other species including the chimpanzee, to communicate with us. If nonhuman species have complex communicative skills, they will surely use these in their natural habitats to communicate with each other, not with us.

Chomsky's argument that human language ability is innate makes language part of the bread on top of the similarity sandwich: a specialized human adaptation. But recognizing that chimps and other apes do not share our system of language does not imply that other species might not have their own specialized systems of communication—part of the bread on their own sandwiches. And the best way to look for these is by going to the places where animals most need to communicate with each other: in their own natural habitats.

One of the nicest studies of spontaneous animal communication started with some observations made by Thomas Struhsaker of Duke University on the vervet monkeys of Kenya's Amboseli National Park in the 1960s. Over many thousands of hours of careful observation and audiotaping, Struhsaker transcribed thirty-six different kinds of vervet sounds. By noting the different situations in which these sounds were made, Struhsaker concluded that twenty-one of them were distinct messages exchanged among the monkeys. His observations formed the raw material for one of the most thorough and systematic studies of a natural communication system in the wild. Dorothy Cheney and Robert Seyfarth, a husband and wife team from the University of Pennsylvania, set out to test whether the vervets' sounds were really a communicative system, and if so, what kind of information the calls carried.

Struhsaker had noted that the vervet monkeys gave distinct acoustic calls to announce three different major predators: leopards, eagles, and snakes. Each alarm call sounds different and produces a different response in the vervet monkeys who hear it.

When a monkey detects a leopard, it utters a loud barking call which prompts other monkeys, to run into trees, where, because of their greater agility, they are difficult for leopards to catch. A monkey produces a short, double-syllable cough when it sees an eagle, and this incites other monkeys to look up into the air, and they may also run into bushes for protection from aerial attack. Snakes evoke a chutter sound from the first monkey to see then, which prompts the other monkeys to stand on their hind legs and peer into long grass. For a human observer, it is hard not to find this bipedal posture endearing, though presumably the vervets themselves are rather stressed by it.

To assess whether these calls were really important communicative signals, the first step Cheney and Seyfarth took was to demonstrate that the monkeys were responding to the acoustic alarm calls and not directly to the predator itself or to other aspects of the behavior of the monkey issuing the alarm call. They achieved this by taping alarm calls and then playing them back to the monkeys when the predator was absent. Sure enough, the monkeys gave the same distinctive response to each alarm call played back on tape as they did to the original call.

Cheney and Seyfarth were able to show that the alarm calls really implied something about the threat (something like, "Watch out, there's a leopard!") rather than the evasive action that should be taken ("Quick, up into the trees!"). They observed that monkeys who were in different locations when the tape with the alarm call on it was played back (on the ground, up a tree, and so on) took different types of evasive action. This indicated that the taped call communicated the threat and not the evasive action to be taken.

Cheney and Seyfarth wanted to go further in understanding how much information the monkey alarm calls contained. They wanted to test whether the monkeys were responding automatically to the sound of the call or whether they responded to the call's meaning. One of Struhsaker's observations suggested a method. Struhsaker had found that the monkeys had two calls with the same meaning. The Amboseli vervet monkeys have two different-sounding calls that warn of the approach of another group of monkeys. One is a long, trilling *wrr* sound, the other a sharp staccato chutter. Did the monkeys understand that these two calls had the same meaning, or did they just respond automatically to the different sounds? To test this, Cheney and Seyfarth developed an experiment out of the story of the boy who cried "Wolf!" once too often. They played back the *wrr* sound when no other group of monkeys was in the vicinity. The first time they did this, the monkeys looked around them for about six seconds. Cheney and Seyfarth played back the *wrr* call again, and again, and again. By the time they had played the *wrr* tape eight times in succession, the monkeys only looked up from what they were doing for two seconds. They had learned that there was no other group approaching, and somebody was just crying "wolf" (or "wrr"). Next, Cheney and Seyfarth switched the call on the tape to the short staccato chutter. If the monkeys were just responding automatically to the sounds, they could be expected to take renewed interest when the sound was changed, and to look around them much longer. But the monkeys responded to the new chutter sound even less than they had reacted to the last of the eight *wrr* sounds. This suggests that they were responding to the meaning of the sound, not just the sound itself.

These vervet monkey communications are termed "referential" because they clearly *refer* to things in the outside world. The monkeys are not just expressing their state of excitement but indicating something about the world around them. The same can be said of the honeybee dance communication system that we discussed in chapter 3: the bees are *referring* to food sources outside the hive. There are other examples of such communication which we shall come to in a moment, but it would be a mistake to assume that all, or even most, animal communications are "referential." In more recent years Cheney, Seyfarth, and their students have been making a detailed study of some of the noises made by the chacma baboons of Botswana. The results do not demonstrate the clearly referential communication that they found in vervet monkeys.

Female baboons utter quiet grunts during a lot of their social interactions. They may grunt as they approach each other, as they feed near one another, as they groom each other, as they move into new areas of their range, and, especially, when they handle each other's infants. These grunts seem to somehow grease social wheels in the female baboon community. Dominant individuals that grunt as they approach subordinate females are less likely to push the subordinate off her foraging spot, and more likely to hold her infant, than dominant females that approach a subordinate silently. Dominant females also sometimes grunt to their victims after they have won a fight. Cheney and Seyfarth used the tape playback methods that they had developed with vervet monkeys to show that subordinate animals that have recently lost a fight with a dominant animal are more likely to tolerate the subsequent approach of that dominant animal if they have heard a taped grunt than if they have not. Though there is some communicative function to these grunts, they do not refer to anything—they are not referential like the vervet monkey calls.

These baboons also make "contact barks." These are loud barks, audible up to a third of a mile away, that female and juvenile baboons emit as they move through forested areas. The chorus of barking baboons sounds for all the world as though the baboons, especially mothers and infants, are calling and answering each other. And this would seem to be a sensible thing to do: one baboon that becomes separated from the group could call out, and then, if its call were answered by a baboon from the main group, it would be able to figure out where its groupmates were. This precisely how the "contact barks" were interpreted by most primatologists until Cheney, Seyfarth, and their colleagues went in with their recording and playback equipment.

Playback experiments showed that female baboons were in fact *not* generally inclined to respond to the calls of other adult females, and their few responses were more likely to come when they were themselves on the periphery of the group. This is rather like one person calling out "Help! I'm lost! Where's the group?" and another answering her, "I don't know—I'm lost too!" Baboons safe in the middle of the pack seemed indifferent to the fate of those who became lost on the periphery. Only the calls from lost infants could evoke any response from the baboons in the main group. Baboon mothers clearly responded to the calls of their lost infants. They would at least look in the direction of the call, and sometimes they would go and retrieve their infants. Never, however, did they respond with a contact call of their own.

Close analysis of the pattern of baboon barks showed that the appearance of calling and answering came about, at least in part, because the baboons didn't tend to utter a single bark on its own: once a baboon started barking, she tended to emit several in succession. Cheney and Seyfarth note that these baboons do not give contact barks in order to share information with their group but as an indication of their own state and position. Indirectly, this can act as a communicative system. If infant baboons call when they are lost, then mother can come and find them. Adult baboons in a group do not go looking for lost adults, nor do they call to them; but at least other lost adults may answer the call. This provides a mechanism for some of the lost baboons to find each other and perhaps raises their chances of together finding the main part of the group.

If it seems surprising that baboons just grunt about their emotional state while vervet monkeys use a referential system of communication, prepare for more surprises. Chimpanzees, though they may have been popular subjects in attempts to teach a human language, turn out to have only a very limited communicative system in the wild.

John Mitani from the University of Michigan and Toshisada Nishida of Kyoto University have studied the "pant-hoots" of wild chimpanzees in the Mahale Mountains of Tanzania for some years. The pant-hoot is a characteristic loud, fairly long (four or five seconds) call that both male and female chimpanzees produce in response to a range of circumstances. It may be uttered in response to another chimp's pant-hoot, or while traveling, or on finding a rich food source or uncovering unfamiliar chimps. The call starts with a series of low pants and builds up through a series of shorter hoots into a climax of one or more high screams. It used to be believed that chimpanzees gave more pant-hoots when they were in large, spread-out groups than when they were in smaller ones and that males who arrived at food sources and pant-hooted would be joined by other chimpanzees. Neither of these contentions has stood up well to closer scrutiny.

At the Mahale Mountains National Park, Mitani and Nishida found that chimpanzees did not pant-hoot when they found food. Pant-hooting was most common when high-ranking males were traveling and had become slightly separated from their companions. Thus it seems that the main function of pant-hooting is to maintain the cohesiveness of male chimpanzee bands. In the volatile societies in which male chimpanzees live, maintaining alliances with other males is centrally important. Thus it seems that pant-hoots are used to enlist the support of buddies. This is an important function—a gang must stick together—but it is not a very elaborate system of communication.

BIRD REFERENCES

The balance so far on referential communication: honeybees and vervet monkeys, yes; chimpanzees and baboons, probably not. How can it be that bees and monkeys outperform their bigger-brained brethren the apes? It seems to run against all reasonable expectations. Well, maybe our expectations are not as reasonable as we think. The resolution of this apparent paradox is that different species have evolved the communicative systems that they need in order to survive and thrive. Honeybees are intensely related to each other and highly social; referential communication serves a purpose for them. Chimpanzees lead more independent lives, and so communication does not seem as important. The communicative skills of different species can only be understood in the context of the lives they lead. We can see this more clearly if we look at a couple of species of birds.

Thomas Bugnyar and colleagues from the University of Vienna recently studied the behavior of common ravens stealing food from the local zoo. These ravens love to swoop in at feeding time and grab tasty morsels intended for the wolves, brown bears, and wild boars. While doing so, Bugnyar noted that the ravens uttered characteristic *haa* calls. These calls served to alert other ravens to the fact that food had just been delivered. Ravens that have found food want other ravens to join them because it is easier to snatch meat away from its intended recipient if there are several of you to distract the rightful owner. By delivering different types of food and recording the calls made by the ravens, Bugnyar and his team were able to demonstrate that the rates of *haa* calls depended on the type of food delivered. The ravens *haa*-ed most to beef, less to leftovers from a local restaurant, and least to wild boar chow. The quantity of food was unimportant. Bugnyar is careful to point out that we cannot be sure that the ravens' calls truly indicate the type of food; the birds may just be more excited when they find more attractive food and *haa* more frequently solely for that reason. But it is at least possible that ravens are emitting referential calls.

An even more substantial body of research points toward the likelihood of referential calling in chickens. Male chickens (older readers may know these as cockerels) like to show off to female chickens (hens) with characteristic food calls when they find food. Chooks also warn of the presence of predators on the ground or in the air. Cockerels do not make these calls for the benefit of other male chickens, so presumably (though this has not yet been proven definitively) males aim to benefit from their calling prowess by gaining sexual access. Improbable as it seems, maybe that is one aspect of chicken behavior that the movie *Chicken Run* actually got right: cockerels really are little lotharios.

Chris and Linda Evans of Macquarie University in Sydney, Australia (another husband and wife team, trained, like Dorothy Cheney and Robert Seyfarth, by the pioneer American ethologist Peter Marler), took recordings of the calls made by male chickens in response to finding food, ground predators, and aerial predators. They then played these tapes back to hens in a test cage. Would a hen understand, on the basis of a disembodied, taped call alone, what the male chicken was trying to communicate? Sure enough, when a food-call recording was played, the hen concentrated on pecking on the ground; when she heard a taped aerial-predator call, she looked upward; and in response to a tape of a ground-predator call, she scanned the ground for danger.

So what is the pattern here? Should we be surprised that honeybees, chickens, vervet monkeys, and maybe ravens produce more referential—in other words, more meaningful—utterances than baboons and chimpanzees? That depends on how you think communication evolved. If communication has been evolving in a progressive march toward the perfection of our own species, then you might expect the degree of complexity of an animal's communicative system to match its relatedness to us. Chimpanzees would communicate more ingeniously than baboons; baboons more than vervet monkeys; vervets more than chickens, and chickens more than honeybees. The evidence clearly refutes such a view.

Instead, we see that where there is a selection pressure for referential communication, it can evolve. Worker bees, who are so closely related to their supersisters and have little chance of having offspring of their own, can communicate three aspects of the outside world (the distance, direction, and quality of a food source) to their hivemates. (Three dimensions is the record for any nonhuman.) Ravens who need each other's help to wrestle meat from intimidating wolves, bears, and boars can call out loudly to their comrades when they find something good to eat. Even gentlemanly cockerels, keen to impress lady hens, pass on warnings about predators and news of tasty food finds. Chimpanzees and baboons, on the other hand, just don't seem to lead the kind of lives where referential communication is all that necessary. Of course, somebody may yet find referential communication in chimpanzees: I'm not suggesting that such a thing is impossible. What I am saying is that if chimps have referential communication, this ability will have nothing to do with their relatedness to us and everything to do with the conditions of their own lives, how they find food and mates and complete the other tasks that evolution demands.

TALK TO ME

Our cat Sybille (whom I had promised would not have her private life revealed in this book) is nudging me as I write to remind me that her feeding time is near. This is a communication she shares promiscuously with all our neighbors. What she does not know is that the tag on the collar round her neck says, "Please don't feed me—I get fed at home."

The traditional view on animal communication, which can be traced back to the seventeenth-century French philosopher René Descartes, was that animals could only express emotions, not thoughts. Descartes wrote, "it is very remarkable that there are none so depraved and stupid, without even excepting idiots, that they cannot arrange different words together, forming of them a statement by which they make known their thoughts; while, on the other hand, there is no other animal, however perfect and fortunately circumstanced it may be, which can do the same." Descartes adds that since many animals succeed in expressing their emotions, they would surely also express their thoughts—that is, if they had any thoughts to express.

Research in the last thirty years has proved Descartes partly wrong, but largely right. Researchers have demonstrated that some nonhuman species can communicate things other than just their own emotional states. It is not clear to me whether we could call these communications "thoughts," but some species certainly do make referential statements: utterances that refer to something in the world around them and are not just indications of their personal state. Sure, a lot of animal communications only indicate something about the communicator: "I'm hungry," "I'm frightened," "I'm horny" are plausible English translations of the utterances of a great many species. But the alarm calls of vervet monkeys, the dances of honeybees, and the crows of cockerels all go beyond this and communicate something about the outside world.

So if Descartes was wrong about vervet monkeys only being able to communicate their feelings, is it possible that these animals (and chickens, and bees) are communicating thoughts in some sense? The leading researchers on animal language, Dorothy Cheney and Robert Seyfarth, have obviously pondered long and hard what vervet monkey communication does and does not tell us about vervet thought. They conclude that the basis of vervet monkey communication is an ability to link a sound (an alarm call or other noise made shortly before a predator appears) with a consequence (an attack by a predator). This is the form of learning we discussed in chapter 3, in which an animal learns about the relationship between signals and the events they signify. It has been found in all the species that have been tested. If the novel sound comes from another monkey, then that's just fine: we have communication. The more difficult side of the equation is to get a vervet to learn to produce a novel alarm call. It seems that vervets are born with a repertoire of sounds and a tendency to emit these sounds in specific circumstances. Though vervets could learn to respond to a novel call through a simple process of associative learning, they would find coming up with a novel warning sound outside the range of sounds that vervets are spontaneously inclined to produce far more challenging. As Cheney and Seyfarth put it, "One suspects that many vervets—even many generations of vervets—would die before a new call, signaling a novel escape response, could be incorporated into their repertoire." Furthermore, there is very little to suggest that a vervet monkey, or any other species of animal except the human, pays any attention to what the audience already knows when it utters a call. Many birds and mammals are more likely to emit alarm calls when friends or relatives are nearby, but they do not distinguish whether or not those friends or kin are ignorant of the threat or already aware of it.

There is little in the communication of animals in the wild, even the referential alarm calls of the vervet monkeys or the dance of the honeybee, that indicates that these animals are thinking anything when they communicate. Their communicative utterances are ballistic responses to circumstances. Others respond to their calls through the simple process psychologists call associative learning, the same way they learn to respond to any other signals or predictable patterns in the environment around them.

The so-called ape language studies, where Washoe, Nim, Kanzi, and the rest were trained to use hand gestures or key presses to communicate, are entirely consistent with the studies of animal communication in the wild. Although the number of signs that these apes learned to use is larger than the number of communicative utterances observed in any species of primate in the wild, functionally the use of these signs was very similar to the natural communications of wild primates. Typically, the signs were used one at a time, to express wants and needs.

Though the term "language" gets bandied about in a fairly

metaphorical sense, it is important to keep in mind what our special human form of communication has going for it. All human languages are constructed of a large but finite set of units (words), but the rules for combining these units permit an effectively unlimited universe of possible communications. Language permits discourse on any topic—not just objects that are present now but things that have existed in the past, may exist in the future, or are figments of our imaginations. Language ensures that there is no ambiguity about the relationship between objects: we do not just name things; we identify their interrelationships. A chimp with a lexicon of one hundred signs or keys to press should be able to say, "The banana you gave me yesterday was tasty." Yet no chimp ever has.

Gary Larson is one of the most intriguing American satirists of the late twentieth century. Quite where his sharp insights into the limitations of animal communication come from is a mystery to me. Larson's thoughts on animal language survive in some very witty cartoons. One of the most interesting demonstrates how Professor Swartzkopf becomes the first human being to understand what dogs are really saying. We see a mad scientist with some funny headgear on and speech bubbles that show us what he hears coming out of the neighborhood dogs' mouths. They all say, "Hey, hey, hey."

Actually, I think Larson is a little too pessimistic. Neighborhood dogs probably recognize each other's barks, and probably each other's state of excitement. Rather than just "Hey, hey, hey," their barks might better be transcribed as, "I'm Rex, and I'm excited." Most dogs respond differently to "Bad dog!" than to "Good dog!" A smart dog responds to "Walkies!" though it may also respond to the word "walk" in the middle of sentences that have nothing to do with any imminent perambulation (like this one). This is an important distinction. Individual words in human language are, in some sense, meaningless. Shark. What does that word mean on its own? Is there a shark here? Will there be one here soon? Have we just missed one? Might I be referring to the sleekness of sharks? Their sharp teeth? Almost anything could be meant—and consequently, next to nothing has actually been said. But a nonhuman always understands a word or call as referring to something in the here and now.

The communication systems in place in nature are quite amazing—I still remember my astonishment when I first heard of the dance "language" of the honeybee—but they are not like human language. What we communicate to each other is not just different from what members of any other species can communicate to their fellows; it is many orders of magnitude more powerful. Other species, even those capable of referential communication, are sealed in worlds in which their thoughts, if they have any, are not uttered. What they communicate is urgent comment on the world around them. They do not care if their audience needs to know this message or not. Some nonhuman species communicate, but the infinitely flexible form of communication we call language, that is the birthright of our species just as echolocation is for bats and ultraviolet vision for honeybees, is simply not available to any other animal.

FURTHER READING

How Monkeys See the World, by Dorothy Cheney and Robert Seyfarth (University of Chicago Press, 1990). One of my favorite books on animal behavior. Cheney and Seyfarth are clearly engaged in their research but not so besotted with their research subjects that they cannot see the limits to vervet communication.

The Language Instinct: How the Mind Creates Language, by Steven Pinker (William Morrow, 1994). Already a classic, this book summarizes our understanding of human language.

Kanzi: The Ape at the Brink of the Human Mind, by Sue Savage-Rumbaugh and Roger Lewin (John Wiley, 1994). Savage-Rumbaugh's account of Kanzi and his capabilities.

Nim, by Herbert Terrace (Knopf, 1979). Terrace's fascinating personal memoir of the experience of raising Nim and what happened after he let Nim go.

Aping Language, by Joel Wallman (Cambridge University Press, 1992). A hard hitting critique of the ape language studies up to 1992.

The Pigeon That Saved a Battalion

On October 27, 1918, the situation of the New York Battalion of the U.S. 77th Division, soon to be known as the "Lost Battalion," was desperate. Major Charles S. Whittlesey realized he had led his men too far into enemy territory around Grand Pré on the Western Front. Nonetheless, he dismissed with derision an invitation from the enemy to surrender: "Come and get us!" he said. For three days his men had repulsed enemy attacks. They were out of food and water, low on ammunition. All attempts to communicate by soldier courier back to their base camp were unsuccessful. Their wires had been cut by the encircling enemy. Three pigeons had been released calling for support, but no help had come. Their only hope was their last pigeon, Cher Ami, one of six hundred pigeons that had been donated by British fanciers to the U.S. Army. Whittlesey tied a message round the bird's leg and watched as it flew off into the smoke of battle. A burst of shrapnel shattered its leg. A bullet pierced the bird's breast. And yet still Cher Ami flew on, covering the twenty-five miles to headquarters in twenty-five minutes. The message capsule was found hanging by threads of ligature where its leg had been. Within hours the 252 surviving soldiers of the New York Battalion (or maybe there were 194 survivors—accounts differ) were safely back with their compatriots.

Cher Ami died June 13, 1919, as a result of battle wounds, but his fame grew and grew. The French awarded him the Croix de Guerre with Palm. He was inducted into the Racing Pigeon Hall of Fame in 1931 and received a gold medal from the Organized Bodies of American Racing Pigeon Fanciers. To this day his stuffed body can be seen at the Smithsonian Institution in Washington, D.C. In 1919 a movie was made about him; in 1926, a poem ("*Mon Cher Ami*—that's my dear friend— / You are the one we'll have to send; / The whole battalion now is lost, / And you must win at any cost"); in 1935, a book.

Today we are not likely to get so excited about a pigeon. And yet perhaps we should. How do pigeons find their way home under difficult circumstances? It's not a trivial question—we couldn't emulate them. Do they have a "sixth sense"? Well, actually yes, and a seventh too. Pigeons have a sensory world quite different from ours. It's interesting that people could get so emotional about Cher Ami as to give him a medal and stuff him for display in the Smithsonian. Our attitudes toward pigeons (and doves, basically the same birds by another name) offer a curious case history in our contradictory attitudes toward animals in general. Pigeons/doves can be heroes, pests, pets, food, sporting goods, demigods. How can one bird occupy so many different niches in our lives?

Battle-hardened soldiers have cried over pigeons; politicians have campaigned for their eradication. Pigeons' ability to home from distant, unfamiliar places, because it is a skill that we don't share, has thoroughly confused some commentators, as we shall see in a moment. But pigeons have been particularly well studied; they are interesting to focus on just because we do in fact know quite a bit about them. Even so, there are still gaps in our understanding of pigeons. These gaps show up the difficulties of studying other species, especially when their perception of the world is so different from ours. Pigeons expose our confused attitudes toward other species and the practical problems confronting a scientific study of animals.

Hero pigeon Cher Ami, who died of wounds sustained in World War I. To this day the pigeon's remains are on public display in the Smithsonian Museum in Washington, D.C. (Smithsonian Institution)

WHEN IS A PIGEON NOT A PIGEON?

Charles Darwin devoted a large chunk of the first chapter of *The Origin of Species* to pigeons: "Many treatises in different languages have been published on pigeons, and some of them are very important." Darwin bred pigeons himself. He collected specimens from far afield, joined the London pigeon clubs, consorted with the fanciers. He knew the names and faces of the different races and marveled at their diversity:

> [C]ompare the English carrier and the short-faced tumbler and see the wonderful difference in beaks. . . . The Runt is a bird of great size. . . . The barb is allied to the carrier, but, instead of a long beak, has a very short and broad one. The pouter has a much elongated body, wings, and legs; and its enormously developed crop, which it glories in inflating, may well excite astonishment and even laughter. The turbit has a short and conical beak, with a line of reversed feathers down the breast; and it has the habit of continually expanding, slightly, the upper part of the esophagus. The Jacobin has the feathers so much reversed along the back of the neck that they form a hood, and it has, proportionally to its size, elongated wing and tail feathers. The trumpeter and laugher, as their names express, utter a very different coo from the other breeds. The fantail has thirty or even forty tail-feathers . . .

Darwin could go on for pages about the wonders of the pigeon. The point of his obsession was that he knew that nobody would deny that the great variety seen in the different races of pigeon was due entirely to human breeders artificially selecting particular desired characters and breeding for them. And yet: "Altogether at least a score of pigeons might be chosen, which, if shown to an ornithologist, and he were told that they were wild birds, would certainly be ranked by him as well-defined species." If human interference could breed such a diversity of types through artificial selection, Darwin argued, surely nature through natural selection could also create a diversity of species by the preservation of better-adapted variants at the expense of those less able to produce offspring in future generations—evolution by natural selection.

English Carrier

Short-faced English Tumbler

English Barb

English Pouter

English Fantail

Rock Pigeon

The many varieties of pigeon as observed by Charles Darwin. Some of Darwin's own pigeons, now stuffed and mounted, can still be seen at his home, Down House, in Kent. (Charles Darwin, *The Variation of Animals and Plants Under Domestication*, 1868)

Darwin knew that all modern breeds of pigeon are descended from the rock dove, *Columba livia*. Rock doves nest on the ledges of cliffs and feed on the seeds of coastal grasses. This lifestyle has prepared them surprisingly well for perching on the window ledges of human habitations and scavenging around bakeries.

We are so confused about pigeons that we give them two different names depending on what we want to do to them. When is a pigeon not a pigeon? When it's a dove. Racing pigeons, for example, are pigeons. Unless they happen to be white, in which case they are called doves and released at weddings. The color of their wings is just an accident of their breeding. Put a male and a female together and you'll soon find that the colors of the wings don't determine what species they are. The same *Columba livia* that's a racing pigeon in the city is a rock dove if it is found feral, nesting on a cliff overhang instead of the ledge of a tall building.

Trafalgar Square is counted by most as the heart of the city of London. Aside from Horatio Nelson atop his column and the four massive lions around the base, the most noticeable thing about the square is the thousands of pigeons. Up to forty thousand birds depositing one ton of feces on Nelson's head each year, by some counts. In 2000 the then newly elected mayor of London, Ken Livingstone, decreed that the pigeons must leave Trafalgar Square. Cleaning up after the birds was costing the city £100,000 a year, Livingstone claimed; their droppings carried influenza, encephalitis, tuberculosis, and other unsavory things. It might seem that the case was simple; after all, Livingstone was only proposing to prohibit the sale of pigeon feed, not shoot the little fellows. But the magnitude of the outcry indicates the depth of feeling about these birds. One of Livingstone's own political allies, Labour M.P. Tony Banks, introduced a motion into the House of Commons urging compassion toward the "gentle London pigeon." Such a motion has no force; it is just an expression of opinion. It merely consumes parliamentary time that could be spent on weightier matters. With the closure of the pigeon food booth at the square, People for the Ethical Treatment of Animals urged supporters to bring their own birdseed to the square. Visitors to their web site were also offered the opportunity to "Click Here to Crap on Ken." (When I tried clicking on this spot—without political motive; solely in the interests of research—it seems all their crap had been used up: all I got was a "404—File Not Found" message.) Another animal rights group, Animal Aid, drew attention to the vulnerability of "juvenile and elderly pigeons."

Why care about the pigeons? Canadian researcher Louis Lefebvre has studied the feeding habits of street pigeons in Montreal over many years and found that these birds never concentrate their grain-gathering activities in one place. Though they may spend the majority of their time in one spot for most of their feeding, city pigeons are always on the lookout for alternative food sources. So, although reducing the availability of food must have some impact on the population numbers, a dramatic decline is unlikely. Animal rights campaigner Carla Lane argued that "if a pigeon lands on a child's shoulder it will paint a good picture in their mind and show them that all animals are worth caring for." Her imagination didn't seem to stretch to the possible consequences if the un-house-trainable bird were to poop on the child.

Could it be that Londoners want pigeons around Trafalgar Square because of a confused folk memory of the role that doves should play around major monuments? Doves make holy places holier; hence the strong association of pigeons or doves with major cathedrals. Think of St. Mark's in Venice. In the Muslim world, doves and pigeons are protected around mosques. Hindus also revere doves. In 1925 two English boys caused a riot in Bombay when they killed some street pigeons. The locals were enraged to see their revered birds harmed. The Stock Exchange and main market closed, and there was a widespread strike.

How did doves get to be holy? From earliest times and in diverse cultures people have seen birds, flying off into the heavens, as important messengers between man and the gods. Small, lighter-colored birds, doves, became charged with religious significance in the ancient Middle East. The dove could carry the person's soul into the spirit world at death. Or carry messages down from the gods in heaven to people on earth. It was a dove that whispered into Mohammed's ear and was his oracle. In Judaism, doves signify the love of God for his chosen people and are also symbols of purity. In Christianity, the dove is the symbol of the Holy Spirit.

Pigeons, on the other hand, are just meat. Before the chicken took over as our major food bird, country houses would keep pigeons/doves in dovecotes that still adorn many a manorial garden. I suppose the reason we eat half a roasted chicken instead of a whole pigeon is because the chickens don't fly away, which makes chickens easier to keep and leads to tenderer meat. But the fact that pigeons could fly off and feed themselves was once considered a bonus. As a handy self-reproducing and self-feeding meat parcel (a pigeon yields about as much meat as a Big Mac), the pigeon is hard to beat. Ancient Mesopotamian inscriptions record that wealthy people were keeping pigeons as a winter protein supply thousands of years ago. When Assurnasirpal, king of Assyria in the ninth century B.C., inaugurated a new palace at Calah, his guests enjoyed at the banquet, among a great many other delights, ten thousand doves and ten thousand sukanunu doves (whatever they were). Nowadays, when restaurants offer pigeon, it is usually labeled "squab," which can denote a young bird of any species. So low has the stock of the pigeon sunk that we won't even eat a beast by that name. Woody Allen called pigeons rats with wings.

Once people started keeping pigeons for meat it couldn't have taken long before they noticed one or two things about these birds. Unlike most animals, pigeons are, to say the least, uninhibited in their courtship and mating. City parks in springtime are full of pigeons and doves cooing, bowing, and mating, quite unabashed by any number of spectators. It is probably this blatant sexuality of the dove that led to its association with the ancient goddesses of love and sex: Ishtar, Astarte, Aphrodite, and Venus. The dove was also sacred to Bacchus, god of wine and revelry. These are not just Western obsessions: the Hindu god of erotic love, Kamadeva, is usually represented with a dove as his steed. Doves made their way to the Virgin Mary through their association with pre-Christian mother goddesses. In the new and puritanical Christian religion, doves kept their association with love, but shorn of its physical, lustful aspects; they became messengers of peace and goodwill.

Pigeons and doves are puzzling, connoting different things to different people. They are familiar to most parts of the world, and yet we tell ourselves such inconsistent stories about them. I cannot find any commentator who stated explicitly the idea that pigeons belong around Trafalgar Square as a way of keeping the monument holy, but perhaps that notion unconsciously lingers in the vehement responses to a relatively modest proposal.

HOMING HISTORY

On rereading Darwin's *The Origin of Species*, I was at first surprised that he omitted to mention the most remarkable human-bred characteristic of modern pigeons, their ability to home. Wild rock doves do not "home" in the way that domestic homing pigeons do. They may forage over a range up to half a mile, but they do not make the epic journeys over hundreds of miles for which homing pigeons are famous. But I had failed to appreciate just what a modern phenomenon the long-distance homing pigeon is. Darwin was writing in the first half of the nineteenth century. At that time, breeders in Belgium were just beginning to select birds for their ability to head for home rapidly over longer distances. Though Charles Dickens commented on the habit in his magazine in 1850, Englishmen knew little about the phenomenon until some Dutch fishermen demonstrated the abilities of pigeons from the Low Countries by releasing 110 Antwerp birds from London Bridge in 1858. In 1858 Darwin was at Sandown on the Isle of Wight (where, many years later, I was to go to high school), condensing his notes into the volume that became *The Origin of Species*. It seems the excitement caused by the Antwerp pigeons either passed him by or came to his attention too late to be included in the manuscript.

It may be hard to imagine in this age of cheap long-distance phone calls and free email, but for years the homing, or carrier, pigeon was the only reliable form of distance communication short of sending a human messenger. Even then, the running human could more readily be intercepted than the flying bird and could not possibly match the pigeon's speed.

It isn't clear when people started exploiting pigeons as message carriers. Ancient texts record that Romans communicated using pigeons. Frontinus gives a fairly detailed account of the use of pigeons at the siege of Mutina (modern Modena in Italy) in 43 B.C.: "Hircius made use of Pidgeons to convey Letters. He shut them up in the dark, and suffer'd them to be very hungry; then did he fasten Letters to their Necks, bound them with a big Hair, and from the nearest place to the Walls of the City he let them fly. They being desirous of Light and Food, mounted on high, to go to the loftiest Houses, where Brutus took them; and by that means he was made acquainted with all Passages; for after that he put Food for the Birds in certain places, and so made the Pidgeons to fly thither."

There is something odd about this early account of pigeons as messengers—something that gives a clue to how the modern homing pigeon came into being. The earliest accounts of pigeons taking messages involve people who were besieged in a castle or city. It seems quite possible that the distances a pigeon would need to fly to carry a message out of a besieged city would not be particularly great, not perhaps further than the bird's normal foraging radius. Most likely the besieged inhabitants of ancient cities kept pigeons primarily for food. The human prisoners in the city doubtless noticed that the birds could fly over the heads of the opposing army. If the birds were not fed at home, that would encourage them to forage widely. They could be let loose with a message attached and the hope would be that the birds would fly over the heads of the encircling enemy to friends on the other side. But this would necessarily be a hit-or-miss affair. Sometimes the pigeons would deliver the message to the enemy instead of to one's friends. Quite often the pigeons would not leave the besieged castle or city at all but instead would hang around the bakery picking up spare grains.

Before too long somebody must have noticed that if a pigeon did make it over the heads of the enemy to supporters coming to the rescue, the birds' return flight would be far more reliable than the outward journey had been. The pigeon's instinct to return home is strong. Returning home the birds were very likely to fly over the heads of the enemy and make a beeline for a specific spot, the home loft. It is this return journey that was very extensively exploited to carry messages in the nineteenth and twentieth centuries.

None of the classical texts mention the possibility of using pigeons to send messages on their return (homing) journeys. Perhaps the authors of the texts, not being pigeon experts themselves, didn't think to record it. Perhaps it was kept a secret and has been lost to history. In any case it is clear that the skill of communicating with pigeons in any form was lost to the West until relearned from the infidels at the siege of Jerusalem in 1099.

So we don't really know when the modern method of using the homing journey of a pigeon to carry a message developed, but it is apparent that by the mid-nineteenth century there was a frenzy of pigeon breeding in northern Europe, with regular competitions, first in Belgium and spreading out from there, to see whose birds could home fastest and furthest.

PARADOXICAL PIGEONS

How can we hope to understand how pigeons home? Some animal abilities are just so far beyond our ken that it seems hard to know where to start. Indeed, the almost magical navigational skills of pigeons have tipped at least one observer over the edge.

The British popularizer of science (or nonscience) Rupert Sheldrake, in his book *Seven Experiments That Could Change the World*, expresses his conviction that the way pigeons home is beyond regular scientific explanation: "The orthodox paradigm [of science] has broken down." What is needed for homing, according to Sheldrake, is "a direct connection between pigeons and their homes." He proposes that "the sense of direction of homing pigeons depends on something rather like an invisible elastic band connecting them to their home, and drawing them back toward it." What can he mean? Sheldrake posits a connection between the pigeon and its home based on the Einstein-Podolsky-Rosen paradox. This is one of those weird puzzles of quantum mechanics by which invisible particles influence each other in ways that appear impossible to laypeople. Undeterred (or perhaps emboldened) by the incomprehensibility of quantum-mechanical concepts, Sheldrake goes on to explain that the quantum quality of "nonlocality" binds the universe into one holistic system. This means, if I follow him correctly, that pigeons can find their home loft even when they don't know where it is. The most interesting thing about Sheldrake's hypothesis is that he posits a specific experiment to test it.

According to Sheldrake, because of the Einstein-Podolsky-Rosen quantum paradox invisible elastic band that connects them to their home loft, pigeons can find home no matter what. In particular, it makes no difference to the pigeon whether you move it from the loft or the loft from it. Sheldrake proposes transferring a bunch of pigeons out of their home loft into a sealed container and then moving the empty loft somewhere else. Ensuring that the pigeons are upwind of their loft (Sheldrake seems here to be acknowledging that the pigeons, given a choice, would prefer to use their sense of smell rather than their invisible quantum elastic bands), the experimenters should then release them from their containers. If Sheldrake is right, the pigeons will find the new loft location.

It is a shame that Sheldrake has not carried out this simple experiment—or chatted with a pigeon fancier about it. For the outcome is a foregone conclusion: any pigeon left behind when the loft departed could never find the loft's new position without an odor or some other cue to go on. In the days when armies kept mobile lofts of pigeons, this was common knowledge. Any birds left behind would never catch up.

Sheldrake is not just wrong in the hypothesis he offers for how pigeons home. He is wrong to imagine that the "orthodox paradigm" of science has been stretched beyond its breaking point by the puzzle of pigeon homing. In fact we have a fair understanding of how pigeons find their way back to their home lofts. They have a sensory world that includes elements that are alien to us, and they combine these senses in ways that are ingenious and still not entirely understood. The problem is not yet completely solved, but

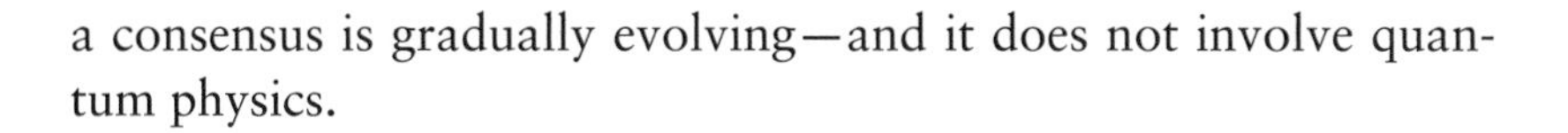

a consensus is gradually evolving—and it does not involve quantum physics.

UNLEASH THE PIGEONS OF WAR

The abilities of the homing pigeon first really broke into the public consciousness when the Prussian armies besieged Paris in 1870–71. There are two problems in using homing pigeons to carry messages, both of which were solved by the French in 1870. The first problem is getting the birds out of your city, over the heads of your besieging enemy, and into the hands of your friends so that they can be released to carry messages back to you. This is the part of the process that the ancients had round the wrong way, which must have led to a great loss of birds and messages. The French solved this problem by taking the birds out of the city in hot air balloons. Though the besieged Parisians were able to take messages out of Paris by hot air balloon, the only attempt at a return flight was unsuccessful: balloons could not be steered with sufficient precision to guarantee that they would return to the city rather than fall into the hands of the enemy. The pigeons, however, homed with unerring accuracy.

The second problem with the homing pigeon is that only a relatively short message could be tied around its foot. How to increase the carrying capacity of the bird without putting such a weight round its ankle that it could not fly? As well as being enthusiastic pioneers of aviation, the French were also pioneers of photography (indeed, the invention of photographic film and moving pictures are both generally credited to the French). A pharmacist in Tours (where the French government had moved as the Prussians approached Paris) had the idea of using photography to reduce messages to the size of a microscopic dot. Photographers shrank messages fortyfold, to the point where they could be conveyed on little films weighing only one-hundredth of an ounce. On receipt the images were projected onto a screen with a magic lantern and copied down by scribes. In this way a pigeon could carry approximately three thousand short messages on each flight. So great was the capacity of the pigeon post that the French authorities soon opened it to personal correspondence. By November 1870, it was even possible to send messages to Paris from England by "Pigeons Voyageurs." Letters had to be written entirely in clear French and sent by registered mail to Tours, where they were photographed, together with many other messages, reduced in size to a dot, and dispatched by pigeon.

In all, during the siege of 1870–71, some three hundred pigeons were released with messages for Paris, and about fifty-nine succeeded in returning to their home lofts. Given that the siege took place in winter, this is not a bad success rate. After the war the French erected a grand monument to the heroic pigeons that had kept Paris connected to the outside world. The next time the German armies reached Paris they melted the monument down.

The siege of Paris made clear that pigeons really were terrific messengers. The breeding of improved homers continued apace. By the outbreak of World War I the major European combatants all had special pigeon divisions (the United States lagged behind and had to borrow British birds when it entered the war). The Germans ordered the destruction of all alien pigeons in the lands they occupied. Notwithstanding the terrible human carnage of that war, hardened soldiers could be prompted to wax poetical on the value of the pigeon. Here is the tribute of one British general, Major General Fowler, chief of the Department of Signals and Communication, to the noble pigeon:

> If it became necessary immediately to discard every line and method of communications used on the front, except one, and it were left to me to select that one method, I should unhesitatingly choose the pigeons.
>
> It is the pigeons on which we must and do depend when every other method fails. . . . when the battle rages and everything gives way to barrage and machine gun fire, to say nothing of gas attacks and bombing, it is TO THE PIGEON THAT WE GO FOR SUCCOR.
>
> When troops are lost, or surrounded in the mazes on the front, or are advancing and get beyond the known localities, then we depend absolutely on the pigeon for our communications. Regular methods

> in such cases are worthless and it is at just such times that we need most, messengers that we can rely on. In the pigeons we have them. I am glad to say they have never failed us.

After the War, massive monuments to the heroism of pigeons were erected in Brussels, Belgium, and Lille, France.

In the Second World War too, pigeons were centrally important to communications. Paratroopers were dropped with pigeons they could release to convey back their safe landing. Bomber crews were given pigeons they could use to relate their position back to base if they were brought down over enemy territory. Many hundreds of thousands, perhaps millions, of pigeons served in World War II. Thirty-two pigeons were awarded the Dicken medal for valor. It was Gustav, carried by a Reuter's war correspondent on to the D-Day beaches, that brought the first news back to Britain of the success of the landings. Kenley Lass brought back important information to Britain from a secret agent in occupied France. An unnamed hero known only by his army tag number, DD.43.Q.879, trained by the Australian Voluntary Pigeon Service, saved a patrol of U.S. Marines whose radio had been smashed while under attack by the Japanese. Julius Caesar was parachuted north of Rome prior to the allied invasion of Italy and returned across three hundred miles of Mediterranean to report back important information on the position of Italian defenses. After the war, many of these birds were stuffed and displayed in museums. Perhaps they still linger on in museum basements around the world.

The only significant development that military use seems to have added to the pigeon's repertoire is the breeding of birds happy to home to a mobile loft. This solves the chief drawback of the pigeon as a means of communication, the fact that, in the normal run of things, the birds will only return to one point, their native loft. Military pigeon trainers developed birds that could adapt to the movement of a loft, and in the Second World War most pigeons worked out of lofts that were trailers drawn by motor vehicles. Of course, the birds needed to be given time to accustom themselves to the loft's new position, and it had to stay still once the birds were released on a mission.

World War II. U.S. paratroopers about to jump with pigeons strapped in corsets across their chests. (Levi, Wendel M., *The Pigeon*, 1963, Levi Publishing Co. Ltd. Photo by Russ Erwin, Grantland Rice Sportlight News)

By the 1950s more reliable and encrypted radio communication had made the pigeons redundant, and, to my knowledge, the last military pigeons (those of the Swiss) were demobilized in 1994. But though their fighting days may be over, fanciers still race pigeons. Belgium remains the number 1 nation for pigeon breeding. The most fancied birds can fetch twenty-five thousand dollars, and the biggest races attract similar sums in prize money. In Britain the annual British Homing World Show attracts around thirty thousand people and raises over one hundred thousand pounds for charity. Queen Elisabeth II, a patron of the society, donated two prize birds from her own loft to their charity auction last year.

Even the World Wide Web has used pigeons as a carrier of information. In 1990 one nerd wrote a protocol specifying how to use pigeons to transfer Internet information. In 2001, a group of pigeon enthusiasts in Bergen, Norway, succeeded in implementing this Carrier Pigeon Internet Protocol (CPIP). Packets of Internet information were carried on the legs of pigeons. The pigeon was 5 trillion times slower than optical Internet cable—a whole web page would have taken about five hours. Asked why he had bothered, a participant offered, "Because no one had done it before."

So how do pigeons home? The first step toward answering this riddle comes, not from contemplating quantum paradoxes, but from looking carefully at the sensory abilities of these birds.

Much of what we now know about the sensory abilities of pigeons stems from an attempt to adopt pigeons for wartime use far less successful than their role as message carriers. April 1940 found the Nazis in Norway and Denmark and the famous behaviorist B. F. Skinner on a train from Minneapolis to a conference in Chicago. Watching a flock of birds lifting and wheeling over the fields, and pondering the recent aerial bombardment of Warsaw, he wondered if the birds' excellent vision and maneuverability could be put to use steering an antiaircraft missile. After returning to Minneapolis a few days later, Skinner bought a few pigeons from a poultry store that supplied Chinese restaurants and began experimenting with ways to get them to operate levers that might control a missile. Over the next few years Skinner gradually perfected a system by which a pigeon pilot could steer an aircraft. The pigeon was held hanging in a man's sock with the toe missing (the sock's, not the pigeon's). In front and beneath it the bird saw a photograph of a scene with a target clearly visible. Between the pigeon and the picture was a cone with four segments. Each segment had a mechanical device attached that could pick up the bird's pecks and, through a system of levers and pulleys, use them to move the scene beneath. Pecks on the corresponding segment of the cone would move the scene back and forth, left and right, as if the "missile" had moved in that direction—the pigeon was piloting the craft. During training, the pigeon would periodically be rewarded with food grains dropped onto the cone in front of it. Skinner was convinced that the pigeon system would work. And furthermore, it was far less bulky than the U.S. Army's first attempts at electronic missile guidance systems. To Skinner's bitter disappointment, however, the army top brass were of the opinion that pigeons were passé; electronics were the way of the future. His pigeons never got to guide a missile in anger.

Though Skinner's pigeons never flew except by flapping their own wings, Skinner and the students who worked with him on Project Pigeon (as a junior naval officer pointed out, not a very clever code name for a project involving pigeons) developed considerable expertise in training pigeons in behavioral experiments. The "missile" cone became the forerunner of a standardized pigeon testing chamber now known as a Skinner box. In the Skinner box, a pigeon (no longer hanging in a sock but standing on its own two feet) is offered one or more plastic keys to peck on, and responses to these keys can be rewarded with food grains through an automated hopper mechanism. By connecting the pecking key and the reward mechanism through a computer, the pigeon's responses can be rewarded however one wants. It is to Skinner's invention, born out of a thwarted desire to be of use to his nation in wartime, that we owe much of what we know about pigeons.

FLY AWAY, PIGEON, FLY AWAY HOME

Imagine you are shut up in a box, the box is placed on a truck or train, and you are transported scores—hundreds—of miles from home, you know not where. Then you are manhandled out of the box into the desert sun and let loose to find your way home. Would you succeed? As a handful of tourists in the Australian outback demonstrates every year—no way! People who stray from main roads in unfamiliar terrain without signposts or telephones are hopeless at finding their way back home. And yet pigeons routinely home under these conditions. How do pigeons achieve what we almost never can?

The pigeon's ability to home seems nearly magical, but it is built on senses that are not supernatural, some of which we share and some we don't.

In chapter 2, when discussing honeybee navigation, I mentioned that many species are sensitive to the time of day. Pigeons are among these species. They use this information, together with the position of the sun, to perform the old Boy Scout's trick of getting compass directions—(also described in chapter 2). We know this from simple experiments that are equivalent to getting pigeons jet-lagged. If the lights in a bird's home cage are switched on and off six hours too early and the bird is released on a homing flight, it will choose a heading ninety degrees from the correct compass bearing. Six hours is one-quarter of a whole day, and ninety degrees is one-quarter of a full circle: the pigeons are making exactly the error a Boy Scout would make if his watch were set to a time zone six hours from where he finds himself.

Pigeons, again like honeybees, can successfully home on days when the sun is obscured. To achieve this, they use a sensitivity to the polarization of light. How could it be demonstrated that pigeons "see" an aspect of light invisible to us? Juan Delius, with Jackie Emmerton and other colleagues at the University of the Ruhr in Bochum, Germany, put pigeons one at a time in a special Skinner box with plastic pecking keys on each of four walls. The pigeons' task was to peck on the keys that were in line with the direction of polarization of the light set into the roof. On different tests the polarization of the light would be altered at random so that first one and then another key would be rewarded if the pigeon pecked it. Delius and Emmerton found that the pigeons solved this task without difficulty, thereby demonstrating a sensitivity to the polarization of light.

The polarization present in any patch of blue sky is especially strong in the wavelengths we call ultraviolet. Could pigeons be sensitive to ultraviolet? Monika Remy and Jackie Emmerton, working with Juan Delius at the University of the Ruhr, put pigeons in a Skinner box where the birds learned to make a response *only* if they saw a light flashed. Remy and Emmerton flashed light of different frequencies at the birds to see what would happen. Just as we would, the birds made the response when "visible" light was shone on them. But they also responded when ultraviolet light, of shorter wavelength than the human eye can see, was flashed at them.

So combine the sensitivities to UV and to the polarization of sunlight and you have a very powerful way of finding the position of the sun. Add to that an innate sense of time of day and pigeons have a reliable compass, at least for days where the sun is visible in some part of the sky.

Bochum, Germany, deep in the industrial Ruhr valley, is not a part of the world where the sun is always visible, and yet pigeons there home even on completely overcast days, when neither the sun nor even a patch of blue sky is available to them. How do they do it?

Maybe pigeons have a magnetic compass in their heads. Fanciers have known for many years that magnetic storms disrupt pigeon homing, and pigeons also have trouble finding home when there is a lot of sunspot activity (sunspots disrupt the earth's magnetic field). Several researchers have tried strapping a little magnet to a pigeon's head to see if its homing would be disrupted, but the results of these studies are not clear-cut. The extent to which pigeons were disoriented by having magnets strapped to them seems to vary a lot from loft to loft and even country to country. Charles Walcott, at Cornell University in Ithaca, New York, was able to demonstrate that one strain of pigeons was confused by a magnetic anomaly in the earth that had no effect on the homing abilities of another strain of birds.

A compass of whatever kind is necessary for successful navigation, but it is nowhere near sufficient. To use a compass you need to know the lay of the land: are you north, south, east, or west of where you want to be? When released on race days, pigeons are not told which way they need to head; that is something they have to figure out for themselves. Once pigeons are within a small radius of home, they recognize familiar landmarks visually and navigate on that basis. But how, when released from a point four hundred miles from home, does a pigeon find the correct bearing, so that familiar landmarks will eventually come back into view? That is the biggest question in pigeon homing, and we still do not have an entirely satisfactory answer to it.

Many European researchers are convinced that pigeons rely on their excellent sense of smell to find home when released from distant, unfamiliar sites. American researchers are much less willing to accept this hypothesis. I can understand how this controversy might come about. Most European pigeon researchers are based in Italy or Germany. Both these nations have a noble tradition of distinctive regional cooking. Think of bolognese or napolitana sauce on spaghetti, salami Toscana, Nuremberg sausages, and Berliner doughnuts. Travel in the United States is not such a gustatory delight. Almost every city has the same McDonald's, KFC, Wendy's, and all the rest. A pigeon trained in Naples Italy, if it had a sensitive nose, could follow the distant aroma of spaghetti alla napolitana until it reached home. One trained in Durham, North Carolina, would find the same fast-food odors tugging it in every direction.

The idea that pigeons follow their noses home gains support from experiments that have compared the homing abilities of two groups of pigeons. One group was brought up in cages exposed to the winds around them; the other group was raised in air-conditioned quarters, protected from the odors of the outside world. When these two groups of pigeons were taken out to test their homing ability, only those that had lived in an aviary open to the winds around them were able to home successfully. The pigeons from an air-conditioned loft were unable to find home. The failure of birds that have had their noses blocked with local anesthetic also supports the idea that pigeons have an odor map that they use for homing. Though these experiments strongly support the hypothesis that smell plays a large role in pigeon homing, contradictory studies, where some pigeons with blocked noses still got home, make this an ongoing debate.

Now this kind of situation—conflicting experimental results—is pretty normal in science. Maybe different breeds of pigeons rely on different cues to different extents. Charles Walcott demonstrated that some breeds of pigeons rely on magnetic cues more than others, so why shouldn't some groups of pigeons depend on odor cues more than others? This kind of unresolved question gets the antagonists very excited, but for those of us with the opportunity to take a longer view, there is no need to see this as anything other than healthy debate. My best guess is that pigeons have a number of navigational systems at their disposal. Which ones they use on any specific occasion may depend on their breed, their experience of what has been effective in the past, and the signals that are particularly salient that day. This is a puzzle right now, perhaps, but not ultimately insoluble. Certainly not grounds for giving up on the immensely successful "orthodox paradigm" of science and grasping for quantum paradoxes and invisible rubber bands, as Sheldrake exhorts us to do.

To most people, pigeons are not the most beautiful animals, and they do not direct themselves toward us in a way that is especially gratifying (unlike, say, cats and dogs, and many other mammals if brought up in human company from birth). But they have a certain self-sufficient elegance and forthrightness of purpose. And, though it may just be an accident of their easy availability and ready trainability, psychologists know more about pigeons than about any other species of bird. To some extent, what has been discovered about pigeons could stand proxy for the wonders we might discover about any other bird if only the requisite time and energy were committed to them.

As a society we may be losing interest in pigeons. It's hard to imagine a pigeon ever getting a medal again or ending up stuffed in a museum. But I'm not sorry we no longer take them to war with us. As Benjamin Franklin said, "Kill no more pigeons than you can eat."

FURTHER READING

The Pigeon, by Wendell Levi, 2d ed. (Wendell Levi, 1981). Originally published in 1957 but still in print and still The Book on pigeons, with fascinating material on every aspect of the human-pigeon relationship. Levi covers the history of pigeons, how to keep them and breed them, pigeon shows, and pigeon homing in a good style and with copious illustrations. Only the section on medical care for sick pigeons is dated.

A Fancy for Pigeons, by Jack Kligerman (Hawthorn Books, 1978). Another somewhat out-of-date but still very interesting book on pigeons and the keeping of them.

7

Monkey See, Monkey Do?

I'm no classical scholar, but it seems to me that the ancients never hit on a major potential difference between humans and other species: the capability to comprehend the implications of what others do.

If I see you stick your finger into an electric socket and recoil in pain, I may choose to refrain from sticking my own finger into electric sockets. I see, I learn. But suppose I see you hiding a stash of cash under the mattress. What I choose to do with that knowledge may depend on whether I think you saw me watching you stash your money. If I think that you don't know that I know that you keep money under the mattress, I might steal it from you (it's not like we're friends). On the other hand, if I think you *do* know that I saw you hiding your money, I definitely won't take it from you: you'd know who to come after, and you're bigger than me. So how I act depends on what I infer about what you know. This ability to act on the basis of the contents of the mind of another being is called *theory of mind*. You and I have it—but could any other species?

How much of what your dog sees you doing around the house

does it appear to comprehend? There was a traditional view in early animal psychology that animals could not understand anything of what they saw another beast do. Latterly there has been a reaction to this view: clearly animals do understand at least some of what they see. Like most fashions in science, the pendulum has now swung way over in the other direction. Now we have primatologists, like Frans de Waal at the Yerkes Regional Primate Center in Atlanta, who believe that animals (primates, at least) learn from each other much as apprentices learn from their masters. De Waal's latest book on imitation and culture in animals is titled *The Ape and the Sushi Master* explicitly to draw attention to what he believes is a commonality between the transmission of culture in people and other apes. According to de Waal, the manner in which bonobos and some other primates learn from each other is very similar to the long apprenticeship that Sushi chefs must go through before they become Sushi masters themselves. De Waal is not the only commentator who believes that nonhuman primates act on the basis of what they believe others know—that they have a theory of mind. I remain skeptical.

Let's go one step further. I do not just have a theory of your mind; I know I have a theory of your mind. I may not have the deep self-knowledge that yogis aspire to, but I am somewhat acquainted with myself: I do have some theory of my own mind. I am self-aware. Is it conceivable that any other species could be self-aware in this sense? And how would we know if they are? Short of the publication of *Confessions of a Dog*, this might seem an imponderable question. Not so—at least according to some researchers. Experiments with mirrors (and not a little smoke) are claimed to show that chimpanzees and maybe a couple of other species are self-aware.

THE MYTH OF THE HUNDREDTH MONKEY

To be able to copy each other is immensely important. How much of what you know now did you figure out for yourself? In reality, most of what we know we learned by instruction or imitation, from modern technological skills like how to drive a car or operate a computer all the way back to basics that even the earliest humans did, like greeting our fellow men and burying them when they die. And so much in between: cooking; making love (or war); making up; nursing babies. Put it another way: Have you ever had an original idea? Come up with a truly original recipe, for example? Why do we buy new cookery books? Did you seriously expect *How to Eat*, by Nigella Lawson ("Britain's funniest and sexiest food writer," according to *Vogue* magazine) to contain recipes you couldn't find in duller books by less photogenic chefs?

I remember being told as an undergraduate that a genius was somebody who had two truly original ideas in a lifetime. This seemed a pretty stupid definition at the time, but I've thought about the problem of originality a bit harder since then. I realize now that aside from some habits that we acquire by direct interaction with the world around us, and a certain amount of trial-and-error fumbling, we are a lot less original than we would like to think. The most "original" thinkers are often people who take ideas that are common knowledge in one domain and reapply them in an area where they appear very novel. The secret of human culture is this psychological action at a distance—this ability to learn from others.

So it's very natural to ask whether this ability to comprehend by observation is shared by other species. How much of what they see (or sense in other ways—it isn't the sense of vision that is critical here) do animals comprehend? We naturally, anthropomorphically, assume that our pets comprehend what they follow with their eyes. With dogs this works up to a point; with cats, far less so.

One of the most celebrated accounts of primate imitation in the wild is the story of the Japanese macaques of Koshima Island. In the late 1940s Japanese researchers started provisioning the macaques with potatoes in order to encourage them to accept human company. This worked well, and from the 1950s onward the macaques allowed themselves to be observed closely. In September 1953 it was noticed for the first time that one of the macaques, a juvenile female known as Imo, took a potato down to a stream to wash the sand and dirt from it. Later other juveniles were observed to wash potatoes in the same way. Then Imo's mother started washing her potatoes too. Over about five years most of the younger macaques picked up the habit. Today, although potato provisioning has all but stopped, on the rare occasions that potatoes are provided, all the macaques on the island immerse them in water before eating.

It must surely be a magical experience to see the macaques of Koshima wash potatoes. I have only seen photographs, but even in pictures the beauty and ingenuity of the animals shines through. The southern Japanese island itself looks wonderful too. The combination of the beauty of the place and the fascination of the monkeys' behavior must be a potent mix. So potent, indeed, that the New Age author Lyall Watson started a bizarre story that once one hundred monkeys on Koshima had mastered potato washing, the skill magically appeared in all the other macaques on the island and on other islands too. The self-proclaimed guru Ken Keyes Jr. has taken Watson's myth of the hundredth monkey and made a parable of it: if only enough people support an idea, that belief will begin to spread to other minds by some supernatural "leap of consciousness," without the need for any of the usual means by which ideas are distributed from person to person.

There is simply no truth to Watson's story. As a comment about macaque society, it is null and void. But judging by the sales of Watson's and Keyes's books, there is a moral to be gleaned here. The true lesson from the myth of the hundredth monkey is what it says about human gullibility with regard to animals. People would rather believe crazy supernatural explanations of animal behavior than straightforward factual ones.

Leaving the supernatural to one side, there is, in any case, a real debate about how these macaques learned to wash potatoes. The standard account, defended by Frans de Waal among others, is that all the macaques after Imo learned by imitating a macaque that had already picked up the idea. In favor of this view is the fact that the first macaques observed to wash potatoes after Imo were her peers and her mother. They, it is argued, had the best opportunities to see what Imo was doing.

In 1990 Canadian psychologist Bennett Galef proposed an alternative view: each macaque learned on its own how to wash potatoes. Perhaps they saw Imo take potatoes to the water, and this may have encouraged them to go down to the water with potatoes themselves, but they did not comprehend what she was doing and did not directly imitate it.

Galef made an interesting point. Habits in a social community are like diseases: the more people have them, the more people get them. If you get on a 747 in London with one person who has a cold, by the time you step off in Perth, Australia, some twenty-four hours later, your odds of having caught that cold are pretty slim: perhaps five or six people sitting near that person will have become infected. But, at the same rate of transmission, if you step on a plane with one hundred people who have colds, then the odds are that by the time you get off, everybody on the plane will be infected. When few people are infectious, few people get infected and the disease spreads slowly. Later, when more people are infectious, more people will get infected, and the disease will spread much more rapidly. The same "slow at first, rapid later" pattern also applies to habits in a social community. Assuming that people were a bit more talkative on long-haul flights than they actually are, then if one person knew a good new joke on departing London, six people might know it by arrival in Perth. But, just as with the cold, if a hundred people knew the joke on departure, everybody on the plane might know it on arrival.

So which was it with the macaques? Did potato washing start slowly and gradually spread faster and faster, as it would if it were a disease or a social habit? Or did it just plod along, with macaques learning at a fairly constant rate, as it would if they were learning independently of each other? The answer, unfortunately, may be lost in the mists of time. It looks like potato washing started small and then developed at a fairly constant rate for several years. But then nobody collected any data for four years. The next time somebody counted, there were lots more macaques washing potatoes. Had the rate of learning suddenly increased in the way it ought to for a social habit? We just don't know: the critical events took place during the four years that nobody was looking.

De Waal dismisses Galef's argument with frustration as just "armchair speculation." According to de Waal, nobody has a right to comment on the actions of the macaques unless they are willing to go to Japan and study potato washing for themselves. De Waal seems to have badly misunderstood the methods of science. The responsibility lies on those who carry out research, especially people who work on populations of animals that cannot easily be studied by other researchers, to report what they have done and observed in sufficient detail for others to test their own ideas about what was going on. If we all had to visit each other's labs or field sites every time we had an argument about what an animal was doing, our science would grind to a halt.

In any case, it turns out that although Galef's argument casts doubt on the possibility that the potato-washing habit was being transmitted by imitation, other activities of these Japanese macaques do clearly show the slow-first, faster-later pattern expected of social learning. For example, the macaques developed the habit of eating fish that washed up on the beach. For the first two years fewer than half a dozen monkeys had the fish-eating habit. Then, over the course of the following two years, sixty more copied them.

Following his own advice, de Waal recently paid Koshima a visit. According to de Waal, the island macaques today are rarely provisioned with potatoes. He notes that in today's Japan dirty potatoes can no longer be bought, so the macaques, on the rare occasions that they receive potatoes, no longer need to take them to water to wash off sand and dirt. Interestingly, however, the monkeys still dip the potatoes in the ocean (in the years since 1953 the macaques gradually switched from washing potatoes in a stream, as Imo had done at first, to washing them in the ocean). According to de Waal, this is because they prefer the salty flavor that the seawater gives them. De Waal emphasizes how this continuation of a habit long after the death of the entire first generation of potato-washing individuals, and even after the initial motivation for the action has been lost, shows commonalities with our human culture, where we maintain activities for generations after the originator has died.

I'm not so convinced. The momentum in the macaques' habit seems to me to differ in important ways from the transmission of ideas in human culture. It took five years for half the macaques in the group to pick up the potato-washing habit. And now, some years (de Waal doesn't say how many) after the potatoes have been delivered to them clean, they are still washing them. This strikes me as a much slower response to changes in the world around them than you would expect in a human group. I remember the change from dirty to clean potatoes in U.K. supermarkets. It didn't take me years to stop washing the clean ones.

GENIUS OR COPY CAT?

The possibility of animals learning by imitation has long been controversial in psychology. It took a very long time to demonstrate that members of any nonhuman species could imitate what they saw others doing. The pioneering animal psychologist Edward Thorndike, working on his Ph.D. thesis in the closing years of the nineteenth century, tried letting cats watch each other escape from a box. Textbooks show these "puzzle boxes" as elaborate devices specially designed to test a cat's ability to figure out how to operate an internal latch and escape to freedom and a piece of fish. In reality they were just orange crates gerrymandered with string and spare bits of wood. In his experiments on imitation, Thorndike used a box with two compartments separated by some wooden bars. He put a cat that was already familiar with the latch mechanism in the side of the box with the latch, and allowed another cat, one unfamiliar with the latch and how it worked, to watch from the other compartment. The experienced cat escaped cleanly and efficiently. Now it was time for the naïve cat that had been watching to have its turn at the latch. Would it have picked up what to do to escape? The answer was a resounding "No." The observer cat was no better at operating the latch mechanism than a cat that had never been introduced to the box before. For Thorndike's cats, seeing was definitely not believing.

But it turns out that Thorndike's method was itself rather naïve.

The social life of cats is not the relaxed and comfortable existence we might imagine from their shared lives with us. A cat desperate to escape from a box is not in the right frame of mind to observe and learn by calmly watching what the other cat does on the far side of the orange crate. Studies in more recent years on the social lives of cats have indicated that cats do sometimes learn from each other. Kittens do pick up some skills by watching their mothers. A study in the 1920s showed, for example, that kittens' choice of the type of rat to kill is influenced by the type of rat they see their mother hunt down.

It took more than fifty years to produce a clear experimental demonstration of imitation in animals. In the late 1970s, Eberhard Curio, a biologist at the University of the Ruhr in Bochum, Germany, tested European blackbirds (a type of thrush) in an ingenious apparatus of his own devising. Blackbirds are feisty little birds that will gang up on a threatening predator to force it away. It had long been suspected that there was an imitative component to this "mobbing" response. In Curio's experiment one blackbird was shown a stuffed owl. Owls are threatening predators to blackbirds and even a stuffed one was enough to trigger a serious reaction from the first blackbird. A second blackbird could see the first blackbird and a harmless colored plastic bottle. Curio constructed his ingenious apparatus so that, from the perspective of the second blackbird, the first blackbird appeared to be attacking this plastic bottle. Would the second blackbird believe that it should attack the innocuous object in front of it because it seemed as though that was what the first blackbird was attacking? Would the second blackbird copy the first? This is exactly what Curio found. The second blackbird attacked the plastic bottle. Clearly, it could only be doing this because of what the first blackbird was doing: blackbirds would never normally attack plastic bottles. A nice demonstration of experimentally induced imitation.

The macaques of Koshima and the blackbirds of Bochum are about as diverse a pair of vertebrates as we are likely to find. The literature now contains demonstrations of imitation in a wide range of species. Researchers at the University of Cambridge recently reported the first evidence of imitation in fish (guppies in a tank will follow a more experienced fellow guppy to safety). It is very clear in retrospect that the long delay in producing clear evidence of imitative learning in animals was due to the inappropriate methods of early researchers like Thorndike and not a reflection of any intrinsic limitation on the part of the animals themselves.

Simple demonstrations that animals imitate each other, however, leave open a very important question: Why? Why do animals imitate? Sometimes people imitate each other unthinkingly, without consideration of the consequences. Yawning, for example, is spread by a simple infectious process. If I mention yawning often enough in this paragraph, I may be able to get you yawning. Yawn. The infectious spread of yawning is interesting, but it could only be a very weak tool with which to build a culture. Far more important in our lives is the imitation of the actions of another because we understand the implications of what they are doing and want those consequences for ourselves. I see Nigella Lawson's orgasmic rapture on biting into some baked salmon, and I think, Yes, I want that for myself.

I spent my first year in Germany working at the University of the Ruhr (not with Curio, though the people I worked with were quite in awe of the famous Prussian with his dignified bearing). The University of the Ruhr is a massive concrete monstrosity from the 1960s glowering on a hillside: Fourteen towers, connected by elevated walkways and underground roads, in a bombastic style suggesting that the architects were looking forward to tendering for the first space stations. To add to that impression, in the midst of all these towers is an enormous circular structure (the university's main theater) which looks for all the world like a docked spacecraft.

It was in a pub in what had been a rural village before the government planners had imposed the university, that I made my first faltering attempts to socialize in German. At first my new friends were happy to translate (their high-school English being far better than my high-school German), but rapidly the good German beer worked its magic, and their German would get faster and more excited, and breaks for translation became correspondingly scarcer and soon dried up altogether. And yet, the strange thing was, as they laughed loudly at jokes I had absolutely no understanding of, I found myself laughing too. "Hast Du verstanden?" (did you understand?) they would ask, slightly surprised, and I would try to cover my ignorance by latching onto something I thought I had heard. My hopeless reconstructions could cause as much laughter as the original joke. The truth is, their laughter was simply infectious: I had to laugh with them. I imitated their behavior with little idea of what had caused it.

Laughter is one form of "infectious" behavior; yawning is another. In social animals, imitation through this simple, uncomprehending route, infectious imitation or mimicry, may be quite common. Mimicry can be enough to ensure that blackbirds join together to mob a predator, Japanese macaques eat stranded fish, or young cats learn to pounce on rats. But it is quite limited as a learning tool. My uncomprehending smiles and nods at German I didn't understand got me into some embarrassing situations. For imitation to be useful in building culture, individuals must be more selective in what they imitate: they must show more comprehension of what others are doing. How can we tell *why* animals imitate each other? How are we to know how much they understand of what they imitate?

Tom Zentall and Chana Akins are psychologists at the University of Kentucky who have made an interesting study of what might motivate animals to imitate each other. Their subjects are Japanese quail, exotic poultry weighing just four or five ounces. Zentall adopted a method developed by the British psychologist Cecilia Heyes but with a pedigree going back to Thorndike's first attempts to demonstrate imitation in animals.

In Zentall and Akin's version of the task, two quail are placed in an experimental box with a transparent plastic partition between them. One quail, the observer, has nothing to do but stand and watch through the partition while the other quail, the demonstrator, demonstrates one of two things. It may demonstrate pecking on a plastic pecking key, or it may demonstrate hopping onto a metal step. The demonstrator quail has previously been trained to perform these actions by the use of food rewards, and it continues to receive food for pecking or stepping while it is demonstrating. First Zentall and Akins showed, as Heyes had before them with rats, that observing animals would copy what they had seen demonstrators do. Observer quail, when given their chance in the side of the box with the key and the step, were more likely to peck if they had watched pecking and to hop if they had watched hopping. But did they do this because to them the acts of hopping and pecking possessed a simple infectiousness or because they comprehended (at some level) the desirable consequences the demonstrator quail were obtaining?

To answer this question Akins and Zentall trained demonstrator quail using the method of partial reinforcement. Simply put, during training Akins and Zentall failed to reward the demonstrators for many of their pecking or stepping responses. They still rewarded them occasionally, and the pattern of reward was quite unpredictable. In this way they were able to engineer demonstrator quail who, like gamblers at a slot machine sustained by very intermittent payoffs, were happy to continue pecking or stepping for five minutes between rewards. With these less demanding demonstrator quail, Akins and Zentall exposed new groups of observer quail to four different types of demonstrator. The observer might see a demonstrator peck or hop, as before, but this time the demonstrating pecker or hopper might be seen to obtain reward (as before), or it might not receive any reward during the period that it was demonstrating. Would it make any difference to the observers if they saw the demonstrator being rewarded or not? Akins and Zentall found that observing quail were far more likely to imitate the demonstrator if they saw the quail rewarded for its efforts. No reward, no imitation was the rule for the quail. They weren't just laughing stupidly at jokes they didn't understand, as I had; they were choosing to imitate or not depending on whether they saw their fellow quail obtain a desirable consequence.

DO APES APE?

Anecdotes about imitation crop up in the scientific literature from many different species. Dolphins and whales; ducks and songbirds; bees, ants, poultry, and Englishmen with insufficient German. Some of these are more convincing than others. Frans de Waal is impressed by the claim that killer whale calves learn how to capture elephant seals by watching their mothers strand themselves intentionally on a beach alongside a seal and attack it. According to de Waal, this is "[p]erhaps the strongest evidence for teaching" in a nonhuman species. However, on closer inspection this result looks far less impressive. In three years of observation and eighty-eight stranding events, only two seals were ever caught. Most of the time (eighty-one observations), there wasn't even a seal in sight. And in one of those two successful captures, it was the whale calf who launched the attack: maybe mother was learning from her baby!

Nonetheless, a certain amount of copycatting seems pretty common in nature. Go up to a horse and yawn—it might yawn back at you. If you stare at a newspaper on the ground, your dog may come over and stare at it too (then again, it may sit on the newspaper and stare back at you). Many animals, especially when young, especially when out with mother, seem to show some ability to learn by observation. Most of this is probably simple infectious imitation, but Akins and Zentall's quail suggest that at least some animals may assess the consequences they see another animal obtain before deciding whether to imitate it.

But surely primates are different. How else did the verb "to ape" come to mean what it does? Where else could the proverb "monkey see, monkey do" have come from? People have always been struck by the alacrity with which primates imitate each other. Claims that apes ape go back a long way. Darwin noted that monkeys "are well known to be ridiculous mockers." There are stories of orangutans brushing their teeth, chimpanzees applying lipstick and smoking a pipe. Surely primates possess a deeper grasp of why others do what they do and imitate in order to obtain those ends for themselves.

This is clearly what Wolfgang Köhler believed. Köhler's pioneering research, carried out while trapped on Tenerife in the First World War, was introduced in chapter 3. Köhler, in describing his chimps' performance on reasoning problems, naturally assumed that they could comprehend what they saw other chimps or people doing. When describing Sultan's difficulty in understanding that he needed to insert the thinner stick into the thicker one in order to reach the fruit placed two stick-lengths outside the cage, Köhler mentions that he helped Sultan by "putting one finger into the opening of the stick under the animal's nose." Clearly Köhler did this because he assumed that the chimp could comprehend the implications of what was shown to him. Köhler goes on to report, however, that this demonstration had no immediate effect on the chimp's behavior.

Jane Goodall, the English ethologist whose pioneering observations of wild chimpanzees in the Gombe National Park in Tanzania have been continuing for four decades, is "almost certain" that chimpanzee infants learn tool use and many other things by watching their parents and other older individuals. She describes, for example, how Evered, a young male chimpanzee, used leaves as a sponge to soak up water from a hollow in a tree while his sister, Gilka, looked on. The minute Evered was finished, Gilka made a sponge of her own and copied what he had been doing. Another young male chimpanzee, Flint, was observed to copy the "mopping" action that adult chimps use to pick up termites.

Köhler's American contemporary, Robert Yerkes, took a wonderful photograph of a chimp reading a book, apparently in free imitation of his human companion. Frans de Waal describes how chimpanzees at the Yerkes Regional Primate Center in Atlanta, Georgia, developed a peculiar posture for mutual grooming. Each chimp would raise one arm high in the air and clasp the hand that its grooming partner had likewise raised. With their free hand, each chimp would check out the other's armpit. This habit slowly spread to all the other adults in the colony.

So it isn't hard to find reports of apes aping and monkeys doing what monkeys see. But the bigger question remains: do primates imitate each other through a simple mindless process of infection (like yawning), or do they copy others selectively in order to obtain the desirable outcomes that they see their fellow apes obtain? Is it possible that their understanding of what they see others do goes even deeper than that?

When we observe the actions of others of our species, it is argued, we do not just admire what they do and assess whether to imitate them on the basis of whether or not they acquire some desirable outcome by their actions. Rather, we recognize in their actions the expression of another individual with motives and a mental life something like our own. We possess a theory of the minds of other people. In recent years several researchers have attempted to demonstrate that chimpanzees have theories of mind something like those we assume for ourselves. I think it will be difficult to find something in nonhumans that is still a controversial theory of how people react to each other. There are psychologists and philosophers who dispute the whole concept of "theory of mind," even as applied to humans. But since this line of inquiry has led to some interesting tests of how much chimpanzees understand of what they see, let's look at the evidence anyway.

One of the earliest attempts to demonstrate experimentally that a chimpanzee might understand the implications of what it sees another individual do was directly inspired by Köhler's tests, described in chapter 3, where bananas were suspended out of reach of the chimp. Psychologists David Premack and Guy Woodruff at the University of Pennsylvania sat a female chimpanzee, Sarah, in front of a TV screen and showed her brief videos of a person attempting to reach a banana that was hung out of reach from the ceiling. In each video the person was unable to reach the banana. Immediately after viewing the tape, Sarah was given a choice of two photographs: one showed the person successfully using a tool to reach the banana (for example, by standing on a box); the other showed the person being unsuccessful in reaching the banana. Sarah was consistent in her preference for the photograph that showed the person successfully attaining the banana. So Sarah's choice of photograph indicates that she had comprehended the brief video. Perhaps she had developed a theory of the mind of the person reaching for the banana and chose the photograph that indicated the fulfillment of that person's desires. But is this the most parsimonious explanation? It seems at least equally probable that Sarah may have just preferred photographs of people eating bananas to photographs of people not eating bananas. After watching a frustrating movie about someone not getting a banana, it may just be more pleasant to attend to somebody getting a banana at last.

Premack and Woodruff invented another interesting test of how much a chimpanzee can understand of what it sees. A chimp was allowed to watch a human hide food under one of two cups. This person then left the room, and one of two distinctively dressed trainers entered. One of these people was known as the *cooperative* trainer, the other as the *competitive* trainer. The trainers were so named because if the chimp gestured to the baited food cup with the cooperative trainer in the room, the chimp would receive the food from the cup. If, however, the chimp pointed to the baited food cup with the competitive trainer in the room, the food was given to the trainer and the chimp got nothing. To obtain food with the competitive trainer present, the chimp had to point to the *empty* container. This container was then given to the trainer, and the chimp received the container with the food in it.

A chimpanzee trained on this procedure was ultimately successful in obtaining food both on tests with the cooperative trainer and on tests with the competitive trainer. This result has been taken to mean not just that the chimp understood what it saw happening (the baiting of the food cup, the presence of one or other trainer in the room) but that it also understood that the two trainers have minds with certain information and attitudes imbedded in them. It has been taken as evidence for chimp theory of mind.

This is a weight of interpretation that the results of this experiment just will not bear. The basic problem here is that there is no necessity for a theory of mind, or really any comprehension of what the people are up to, to consistently obtain food in this task.

Imagine that, instead of two trainers, we had two lights in the room: say, a red and a green. When the red light comes on, the chimpanzee has to point to the filled cup to obtain food; under the green light, it must point to the empty cup in order to be fed. There can be no theory of mind here, because there are no minds involved, just different-colored lights. And there is no reason to suppose that a chimp would not be able to solve this version of the task. It just demands the kinds of simple associative learning about cues and their consequences that have been found in all animals from wasps to human beings. The chimp learns to associate the red light with pointing to the filled cup to get food, and it associates the green light with pointing to the empty cup to get food.

Consistent with the interpretation that the animal was forming simple associations between cues and consequences in order to obtain food is the observation that its learning was not rapid. It took over one hundred attempts before the chimp selected the appropriate cups consistently in the presence of both the cooperative and competitive trainers. This is much more like the gradual formation of associations than the kind of behavior we would expect if it comprehended the knowledge and attitudes—the minds—of the two trainers.

Premack went on to suggest another experiment to try and answer some of the criticisms of the cooperative/competitive trainer experiment. In this new experiment, the roles of experimenter and chimpanzee were reversed. Now, instead of the chimp choosing a container and pointing to it, humans point to containers. Daniel Povinelli and his colleagues at the New Iberia Research Center in Louisiana put this experiment into practice.

In a situation rather like a strange elaboration of the three-card hustle still performed to fleece punters in major cities, four upturned cups were placed in front of the chimpanzee and covered over so that the chimp could not see which cup food was placed under. One person (the *knower*) placed the food under a cup. Another person (the *guesser*) was not present and therefore could have seen where the food was placed.

The cover was removed from the cups so that the chimpanzee could see them. Each trainer pointed to a cup. Would the chimp select the cup pointed to by the knower, the one with the food under it? Or would it choose the cup pointed to by the so-called guesser, which never covered food? Two of the four chimpanzees tested on this task ultimately learned to choose the cup pointed to by the knower and to ignore the cup chosen by the guesser.

Here again it might be that the chimpanzees (at least the two successful ones) comprehended what they saw and operated with a theory of the minds of the two trainers. But yet again an alternative explanation in terms of simple associative learning is possible. The two trainers may have meant no more to the chimpanzees than if red and green lights indicated the empty and filled food cups. This explanation seems particularly likely in view of the fact that only half the chimpanzees were successful on the task, and these two required over one hundred training exposures before any success was demonstrated.

How to tease apart the competing explanations of the chimps' behavior: associative learning versus true comprehension and a theory of mind? One way is to look at how these animals cope with a change in the procedure. Perhaps the slow learning of the chimpanzees is not due to slow, incremental, associative learning but to difficulties in comprehending what was required of them. Perhaps the one hundred practice runs, rather than stamping in an association between the actions of a particular trainer and the food-containing cup, actually served to illustrate to the chimp what the peculiar goings-on of these odd humans led up to. Once the point of the experiment had been communicated to the chimpanzees, perhaps they did then comprehend what they saw and operated with a theory of the minds of the trainers.

To attempt to address this issue, Povinelli changed the structure of the guesser/knower experiment for a final test. Now, neither the guesser nor the knower baited the cup. Rather, a third trainer performed the baiting operation in the presence of the other two trainers. The new knower watched how the cup was baited. The new guesser had his eyes covered with a bag. The chimp could see everything that was going on except which cup the food was put under, because of the screen between it and the cups. The rest of the experiment (the removal of the screen and the pointing by guesser and knower) was the same as before.

In the course of thirty tests of this new type, three of the four chimpanzees developed a preference for the cup pointed to by the new knower. Though the learning was quicker than it had been on the original task, the performance was not perfect. Nor did it appear spontaneously as a result of their previous training. Is this what we would expect of the possessor of a theory of mind? From an individual that could understand the implications of what it sees? I think not. We might excuse the chimpanzees their slow learning in the original guesser/knower experiment because of the difficulties in explaining to them what they had to do. But in the second phase of the experiment, the demands on the chimps were the same as before. Surely the possessors of a theory of mind should pick up on the second phase quite spontaneously?

It seems far more likely that the chimpanzees in these experiments do not comprehend what they see performed in front of them in terms of the minds of the human actors. Rather, they are gradually making a fair proportion of successful choices by the incremental adoption of associations between certain people and actions and the desired food-reward consequences.

THREE WISE MONKEYS

Povinelli himself originally argued that his experiments indicated that chimpanzees had a theory of mind, but more recently he has become less sure. One very simple but telling experiment seems to have particularly influenced his opinion. Povinelli recognized that the ingenious experiments with guessers and knowers were very demanding on his chimpanzee subjects, and consequently their comprehension of the actions of the people in front of them may have been bogged down in a great deal of procedural coming-and-going that muddied interpretation of the results.

Povinelli noted a very common behavior in his chimpanzee colony and recognized that this could be exploited in an interesting way. The chimpanzees at the New Iberia Research Center were always keen to beg for food from their caretakers. Povinelli and his coworkers realized they could use this begging behavior to test how much the chimps comprehended of what they saw, and whether they had theories of the minds of their human caretakers. In an interesting reversal of the roles of the three wise monkeys (see no evil, hear no evil, speak no evil), Povinelli and colleagues offered their chimpanzees a choice between begging for food from a person who could see them and begging from one who could not. The chimps might be confronted, for example, by a choice between a person with a blindfold over her eyes and one with a blindfold over her mouth, or between a person with a bucket over her head and one holding a bucket next to her head, or between a person with her hands covering her eyes and one with her hands covering her ears. In all cases, only one of the two people could see the chimp, and only the person who could see the chimp would respond to begging by providing food.

To the experimenters' surprise, the chimps were just as likely to beg for food from a person who could not possibly see them as from a person who could see them clearly. With enough experience, the apes could gradually figure out whom they should address for food, but they showed no spontaneous understanding that being unable to see would disqualify somebody from providing treats. Only if the person had her back to them did the chimpanzees spontaneously recognize that she was not worth begging from.

These results are very striking. The exercise exploits a behavior, begging for food, which, for captive chimpanzees at least, seems very natural. It does not require elaborate, time-consuming training. Povinelli's group has been able to test six animals, a larger number than many other studies of chimps' theory of mind. So is it the last word on chimpanzee comprehension of other's motivations? Not yet.

Brian Hare and Michael Tomasello are Americans working at the Max Planck Institute for Evolutionary Anthropology in Leipzig, Germany. They were very struck by Povinelli's failure to get chimpanzees to comprehend the implications of what they saw people doing, because Hare and Tomasello had been able to show this skill in dogs. In tests on ten dogs, half were able to find hidden food by following the gaze of either a person or a dog.

Hare and Tomasello have argued that Povinelli's begging experiments are unnatural because they demand that the chimps cooperate with humans to obtain food. Chimps don't cooperate to obtain food, they fight over it. If a chimpanzee is going to show a theory of mind and an understanding of the implications of what others can see, it is going to be in a context of competition over food, not in a context that asks them to decide whom to address for donations.

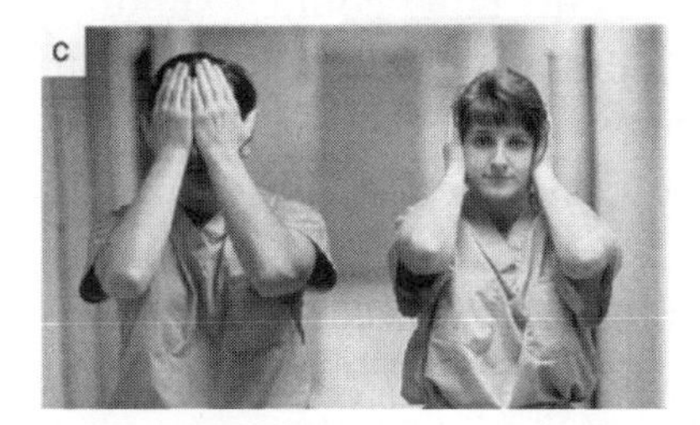

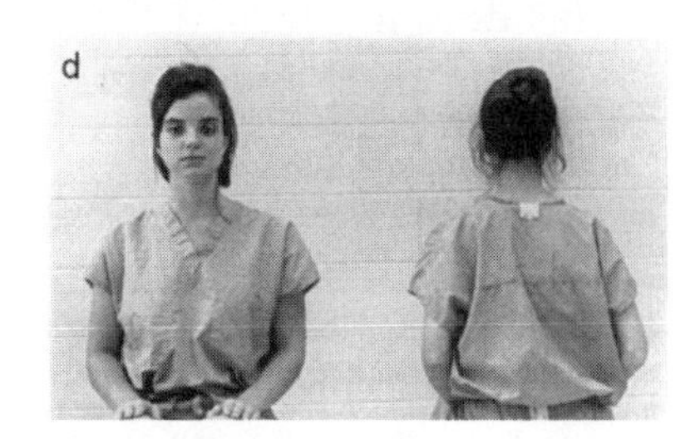

Four pairs of people who presented themselves to Daniel Povinelli's chimps. In each pair, only the person who could see the chimp would provide it with food. Nonetheless the chimps begged for food from either person indiscriminately in all cases except (d). Only in this final case, where one of the assistants had her back to the chimp, could the animal understand that here was someone who was not going to offer it food. (Reprinted by permission of Oxford University Press)

Hare, Tomasello, and their colleagues exploited the information they already had about the dominance patterns in their troupe of captive chimpanzees. Chimpanzee society is strictly stratified; the king of the jungle really does lord it over his underlings. A lower-ranked animal should know never to compete with a higher-ranked one for food—the outcome could only be tears.

The experimenters selected a high-ranking and a low-ranking chimpanzee and placed two pieces of food between them. One piece of food was placed so that both could see it, and, furthermore, each could see that the other could see it. They stared across a room at each other with this bit of food between them. A second piece of food was hidden behind a barrier so that the subordinate animal could see it, but the dominant could not. Furthermore, the subordinate animal could see that the dominant animal could not see it. Would the subordinate animal understand that the dominant animal would go for the piece of food that he could see and leave the subordinate animal in peace with the other food item?

Hare and his colleagues found that subordinate chimpanzees did indeed (and quite spontaneously) prefer the piece of food that the dominant chimp could not see. Subordinate chimps wisely chose not to challenge the dominant animal for the piece of food in view of both of them.

This does not seem a surprising result: How could a subordinate chimpanzee *not* be wary of fighting with a dominant individual for a piece of food? But does this show that the downtrodden lower-ranking chimpanzee has a theory of the mind of its superior? Is it not possible that a subordinate chimp could be afraid of the gaze of a dominant animal without knowing why it has that fear? Surely, facing down the angry stare of a dominant chimp is the kind of thing a subordinate chimp would avoid, whether or not it had a "theory of mind" regarding the dominant chimp.

The proponents of chimpanzee theory of mind will complain that I am being too hard on their noble efforts. I certainly respect how much energy goes into designing and carrying out these studies. But so long as simpler explanations of an animal's behavior are available, the jury must stay out on whether these animals have a theory of mind.

MIRROR, MIRROR, ON THE WALL

What about having a theory of one's *own* mind, being self-aware—could any other species have that? If it is tough uncovering whether other species have a theory of others' minds, how could we ever know whether any nonhumans have a theory of their own minds? Several researchers believe that we can use mirrors to peer into the self-awareness of other species.

Narcissus was not a narcissist. He fell in love with his reflection only because he was very beautiful and did not realize that the face he saw reflected in the pond was his own. If this seems hard to believe (it is, after all, only a myth), consider this report from the

Sir John Ross's men amusing themselves with some Eskimos. Note the Eskimo astonished at his reflection. (J. Ross, A Voyage of Discovery, 1819)

Arctic explorer Sir John Ross on his contact with Eskimos in northern Greenland in the early nineteenth century: "On seeing the faces in the [looking] glasses, their astonishment appeared extreme, and they looked round in silence for a moment at each other and at us." Later the Eskimos were enticed on board ship, where His Majesty's officers and men amused themselves by showing the Eskimos a magnifying mirror: "Their grimaces were highly entertaining, while, like monkeys, they looked first into it, and then behind, in the hopes of finding the monster which was exaggerating their hideous gestures." The Eskimos also thought that Sir John's portrait of his wife was alive. It seems hard to imagine human beings who couldn't recognize themselves in mirrors (especially Eskimos: wouldn't they have seen their reflections in the ice and water often enough?), but the reality is that mirror self-recognition is a learned skill. It develops in children around the second year of life after an undocumented number of occasions in which they are shown their own images in mirrors. It happens that this development of mirror self-recognition occurs around the same time that children start to indicate an understanding of themselves as independent beings.

Most nonhuman species make nothing of what they see in mirrors. Or, if they perceive anything, it is another member of their species. They may attack it or try to make love to it, but they do nothing to suggest that they recognize it as themselves.

Chimpanzees, however, are a striking exception. After an initial period of reacting to their mirror reflection as if it were another chimpanzee, chimps learn that the mirror shows them themselves and use it to inspect areas of their bodies (such as the inside of the mouth and the ano-genital region) that are not normally easily visible to them. Other cases of mirror self-recognition have been reported in orangutans, and hotly disputed claims and counterclaims for this ability have been made for dolphins and a gorilla. No other species tested (and there have been a lot, from fish to dogs and parrots to elephants) has shown any signs of recognizing itself.

But one researcher's "chimp examining its ano-genital region with the aid of a mirror" is another scientist's chimp scratching its rear end because it has lost interest in the odd individual in the looking glass. What was needed was a litmus test for mirror self-recognition.

Gordon Gallup, a psychologist from the State University of New York in Albany, proposed the mark test. It goes like this. While anesthetized, a blob of nonirritating, odorless ink is placed on the ape's eyebrow or ear (areas of the face that cannot be seen without a mirror). The same test can be applied to sleeping children. On waking the individual is shown itself in a mirror. Will it now see the mark on its face and make moves to touch it? Touching the mark on its body is taken as recognition that the face in the mirror is its own. Chimpanzees, orangutans, and (possibly) a gorilla have all passed Gallup's mark test, but gibbons and more than a dozen species of Old and New World monkeys have all failed.

What are we to make of these mirror-using apes? Gordon Gallup and his supporters argue that the mark test proves, not just that certain nonhuman apes recognize themselves in mirrors, but that this self-recognition is evidence of a self-concept. Gallup has argued that passing the mark test indicates the beginnings of something like our human awareness of ourselves—an awareness of the self as an individual entity, even an awareness of its mortality.

I don't believe it for a moment. For one thing, though Gallup and his supporters like to tell a simple story of the ape that woke up and instantly recognized itself in the mirror, the details of the now numerous experiments that have used the mark test do not support Gallup's account.

Nicely controlled studies that have compared the rate at which an ape touches a mark on its forehead or ear with and without a mirror present have found that the presence of a mirror makes much less difference to the animal's behavior than the typical summary of this research suggests. It is not the case that chimpanzees *never* touch the dye mark in the absence of the mirror but touch it energetically as soon as the mirror is introduced. In one of the few studies to report the frequency with which chimps touched their dye marks with and without a mirror, Povinelli and his colleagues reported that on average chimps touched their marks two and a half times in half an hour without a mirror and only just under four times in thirty minutes with a mirror.

According to Gallup, recognizing oneself in a mirror is evidence of self-awareness, and the fact that only humans, chimpanzees, and orangutans pass the mark test shows that this ability evolved before the arrival of modern *Homo sapiens* but sometime after apes had separated off from monkeys. Gorillas present a problem in this neat scheme, since they are more closely related to humans than are orangutans and yet do not appear to pass the mark test. Dolphins too would cause a problem if it could really be demonstrated that dolphins admire themselves in mirrors (I'll come to dolphins in a moment).

But the bigger problem with the excitement over the mark test is the massive tug of anthropomorphism that animates it. "I recognize myself in a mirror and I am self aware," the researchers are saying. "Therefore, if another species recognizes itself in a mirror, it must be self aware too." This argument is fallacious on several levels.

First, imagine you meet somebody who cannot recognize himself in a mirror—would you assume he had no awareness of self? Blind people do not recognize themselves in mirrors, nor do some people with a very rare form of brain damage known as prosopagnosia. (Oliver Sack's *Man Who Mistook His Wife for a Hat* was a prosopagnosic.) And yet nobody ever suggests that these people lack a concept of self. If you talk to a blind person or a prosopagnosic, it soon becomes quite obvious that he possesses a self-concept.

Well, you might reasonably ask, are there any people whose sense of self is in question, so that we might see how they react to a mirror? Indeed there are. Autism is a very wide-ranging syndrome, but there is considerable consensus that autistic people fail to develop the normal sense of themselves as individuals. Their lack of self-concept is measured on tests where their ability to understand other people's perspective and knowledge is explored. Interestingly, although autistic children's self-concept is impaired, their ability to recognize themselves in mirrors develops at a similar rate and in the same way as for normal children.

So we have some people with a normal self-concept who fail to recognize themselves in mirrors (blind people and prosopagnosics) and other people who do not have a normal self-concept but who do recognize themselves in mirrors (autistic people). Shouldn't that be enough to throw out the idea of using mirrors to test for self-concept?

Cecilia Heyes from University College in London has suggested that, in so far as the mark test demonstrates anything, it shows that an animal has what might be called an "own-body" concept: it is able to differentiate between itself and the rest of the world. But, she goes on to point out, the problem with this own-body concept is that surely all animals must have one, not just those few that pass the mark test.

How about a mark test adapted for an animal that depends on smell much more than sight? In a recent study that anybody willing to get her hands dirty could confirm for herself, Marc Bekoff showed that his pet dog Jethro recognized the smell of his own urine. Bekoff went around picking up yellow snow that Jethro and other dogs had urinated on and moving it to spots further up the path where Jethro would come across it. Jethro spent less time sniffing and was much less likely to urinate on top of snow that contained his own urine than snow contaminated with the urine of another dog. Does Jethro have self-awareness based on urine—"I pee; therefore I am"?

Heyes suggests that even to move through the environment without bumping into things implies an own-body concept. So why do some species respond to their bodies when they see them in mirrors and others do not?

Maybe in accepting the mark test as a measure of self-recognition, we haven't tried hard enough to get out of our human perspective on the seeing-oneself-in-a-mirror experience? What would it actually be like if you looked in a mirror and saw something that moved with your every move and was shaped like a human being, but which you did not recognize as yourself? This is a very difficult situation to imagine. I think it would feel something like the actors' exercise which involves pretending to be somebody's mirror. You stand opposite them, and everything they do, you do. It is killingly funny to watch, and very hard to do well.

Imagine, then, that you are confronted with an actor somewhat like yourself (same sex, height, etc.) who effectively mirrors your every action. This would have to be what individuals perceive who have good vision but no comprehension that a mirror is showing them themselves. How would you react? What would you do? And, most pertinently here, how would you react if you saw a spot on the actor's eyebrow or ear? Is it not possible you would touch the same part of your own body? I think you would.

To the best of my knowledge no such experiment has been attempted. But my hunch is that an individual willing to touch her ear on average two and a half times in half an hour when left on her own would touch her ear about four times in half an hour when confronted by an imitating actor whose ear was marked with red paint. This does not mean that she sees the actor as being herself. Just seeing a mark on the face of the person in front of you can be enough to prompt you to touch your own face.

DOLPHINS THROUGH THE LOOKING GLASS

So I see problems with the ready overinterpretation of what apes do in front of mirrors. But at least chimpanzees have hands to touch their faces with.

On May 1, 2001, the *New York Times* ran the headline "Brainy Dolphins Pass the Human 'Mirror' Test." I'm not sure why "Mirror" was put in scare quotes, but the addition of the adjective "Human," to make clear that this is a very human concept, seems fully appropriate. Over the last decade, several researchers have claimed to have shown that dolphins respond to themselves in mirrors, but the report the *New York Times* was referring to, by Diana Reiss and Lori Marino from Columbia University in New York and Emory University in Atlanta respectively, was the first to claim to have carried out Gallup's mark test on dolphins. The problems in administering the mark test to a dolphin would appear insurmountable. Dolphins, because their breathing is under cortical control, cannot be anesthetized (they would stop breathing and die), so how is the mark to be applied without them noticing? Furthermore, dolphins have no hands with which they might touch a mark if they did notice it, and without a clear behavioral reaction to a mark how could any test be carried out?

To get round the problem of not being able to anesthetize the dolphins, Reiss and Marino developed an ingenious "now you see it—now you don't" procedure. On some randomly chosen tests the experimenters marked a part of the dolphin's body that the animal could not usually see with harmless black ink from a marker pen. But on other tests the experimenters went through a charade of "marking" the dolphin's body with a marker pen filled only with water. On these sham-mark tests they went through all the actions as if they were drawing a mark on the dolphin; the only difference was the absence of ink. Nothing in the dolphin's behavior after marking and before approaching a mirror suggested that it could tell whether it had really received an ink mark or not.

This strikes me as an improvement on the method used with apes, where the animal is anesthetized before receiving a mark. Some of the difficulties in interpreting the apes' behavior have arisen from the possibility that the animals were still drowsy from the anesthetic when they were shown the mirror.

But how to cope with the fact that dolphins have no hands? This is a trickier problem. Reiss and Marino's solution was simply to videotape the dolphin's behavior in front of an underwater mirror (not one held above the water as the *New York Times* illustration showed). Assistants who did not know whether the animal had been marked or not scored the behavior of the dolphin in front of the mirror. How much time did the dolphin spend with its body oriented in front of the mirror so that it could see the mark?

In three sham-marked tests administered before any real marks had been applied, one dolphin spent less than a second in front of the mirror. OK—no mark on body, no interest in mirror. Then with a proper mark on a part of the body not usually visible to it, the dolphin spent nearly a minute in front of the mirror, in such a posture that the reflection of the marked area would be visible to it. Mark leads to dramatic increase in interest in looking at self in mirror—so far so good. After the real mark tests had started, further sham markings were performed. Now the dolphin spent an average of forty seconds in front of the mirror. How are we to interpret this substantial increase in mirror self-observation in response to sham markings after the dolphin had had experiences of real markings (from less than one second in front of the mirror before to forty seconds after)? Reiss and Marino argue that this is due to the dolphin now inspecting its body to see whether or not a mark had been applied. They fail to notice that if the only thing we can measure is how long the dolphin spends in front of the mirror, and if the dolphin spends as long looking at itself without a mark as with one, then we have no way of telling whether the dolphin is interested in the mark or not. The test is null and void.

In any case, even if we accept that the dolphins have noticed the mark, does that mean they have done the same thing that apes do when touching a mark only visible in a mirror? In his first report on the mark test, Gallup made much of the reaction of chimpanzees to their fingers after touching the mark. They would look at their fingers and, in one case, sniff them, apparently to see what the mark was. Nothing like that can be assessed with the dolphins. By removing the requirement that the animal actually touch itself in the mirror, Reiss and Marino, though obviously making a necessary adjustment to the dolphin's lack of hands, have simplified the task too substantially.

Let's go back to basics. Why would you touch a spot on your

own face if you saw a spot on a face in a mirror? I believe that aside from recognizing the face in the mirror as your own, there are at least three major components to the answer to this question. First, you have seen that face in the mirror many times before without that new spot; second, the image in the mirror moves in strict synchronization with your own movements; and third, you care about the presence of spots on your face. If you didn't have sharp enough vision to see spots on your face, or the way the image in the mirror moves when you do, you wouldn't touch a spot on your face. If changes in the details of your facial appearance were not important to you, you wouldn't touch a spot on your face. If you had not seen yourself often enough in a mirror to be sure what you were supposed to look like, or didn't have the memory to recollect the details of the changing landscape of your face, you might not touch a spot on your face.

Mirror self-recognition is not an acid test of self-awareness; it is a complex skill built on many component abilities involving acute perception, memory, and the motivation to inspect one's face. Take away any of these important components and an individual will not inspect her face when shown a mirror. Our comfort in our human existence has blinded us to this simple fact. None of us remembers what it was like to learn to recognize our mirror image as ourselves, and few of us will ever have to confront forgetting who that person in the mirror is.

Pointing at a dot on your body when shown a mirror is not some index of humanlike self-awareness (however defined). Rather, it is a skill born of a particular set of sensory abilities and behavioral inclinations. Instead of an unruly rush to be the first to demonstrate self-awareness in another species, what we need are careful studies investigating just what different animals see when they look in a mirror. Unfortunately, very little such work has been done.

WHOSE CULTURE IS IT, ANYWAY?

The ability to understand the implications of what we see others do is one part, an important part, of what gives us culture. Culture makes us what we are as a species. Furthermore, the defining characteristic of our culture are important to the sense of worth of the group we belong to. Culture is the end result of the processes of imitation and instruction that we go through as children and maintain as adults.

To understand why culture is so important to us, consider what a human being without culture would be. A person without culture is an individual in solitary confinement, without even the apparatus of a prison to provide him with food. An individual without the several millennia of culture we have behind us would just be a hairless ape, able to eat nothing but the ripest fruit and vegetables. He would have no shelter, no clothing, no hunting techniques, no meat, no specially bred crops or beasts of burden. He would be naked, puny, and pathetic.

The apes are dying out. Human activity doesn't help, but the great apes have been in decline since before *Homo sapiens* even existed (they have been declining for about 10 million years). So what makes the difference between the several great ape species that hang on in pockets of equatorial Africa and Southeast Asia and the one species, *Homo sapiens,* that has become the most widespread large mammal on the earth's surface? The answer is culture. Our ability to steal ideas from each other.

If you can imitate, even if you don't understand why you copy what you see others do, then you can build a kind of culture. But if you don't comprehend why you are copying what others do, it will be a culture limited in its utility and survival value. You will be like the rather dim-witted, but not evil, child who gets dragged into the orbit of smarter children with less than noble motives. But if you do comprehend motives and outcomes and selectively imitate things that have good payoffs—well, then you are as smart as the smartest among you.

In 1999 *Nature*, possibly the most prestigious of all the scientific journals, published a review coauthored by the leaders of nine chimpanzee field research projects. This article was simply entitled "Cultures in Chimpanzees." The purpose of the review was to document the different habits of different groups of chimps. The chimps of Mahale (Tanzania), for example, pick their noses with sticks—a habit not found in any of the other studied groups. On the other hand, the chimpanzees of Bossou (Guinea), Taï Forest (Ivory Coast), and Gombe (Tanzania) commonly smash food onto wood with their feet to break it up, something the chimpanzees of Mahale never do. In all, the nine authors documented sixty-five behavior patterns that ranged from rare to common among the chimps they studied. This, they argued, was culture: things done by one group and not another.

And quite right too. If it can be shown that these various habits are learned by imitation of, or instruction from, other members of the tribe—and there is no particular reason to suppose that at least some of them are not—then why not call this culture? We are seeing here, as Darwin would have expected, "no fundamental difference" between our own mental life and that of other species, but rather some of the "numberless gradations" he predicted.

But on the other hand—how slight this culture is. In no case can these expert researchers, with their sixty-five behavior patterns from nine chimpanzee societies, point to a single case where the putative cultural activity is essential to an individual chimpanzee's survival. How many culturally transmitted behaviors could be found in nine human societies? The question is crazy: even in a single human society the number of culturally transmitted behaviors is beyond counting. And fluency in these cultural patterns of action is absolutely essential to the survival of the human individual.

In chapter 5 I mentioned Winthrop and Luella Kellogg from the University of Indiana, who in 1931 adopted a seven-month-old chimpanzee. Gua was brought up alongside their son Donald, who was two and a half months older. This is one of the earliest detailed studies of the development of a chimpanzee. Gua was reared for nine months in a manner matching as closely as possible that of Donald, and the two were subjected to regular testing of their psychological faculties. Although the Kelloggs were expecting to see imitation by the chimpanzee, they were surprised to find that the human child was far the greater mimic. Indeed, Donald would even imitate the ape's characteristic food-demanding grunts.

Humans are more inclined to mimic and, as they grow older and wiser, more selective in what and who they imitate, than even chimpanzees. In a recent study, two Japanese researchers, Masako

Myowa-Yamakoshi and Tetsuro Matsuzawa of Kyoto University, demonstrated forty-eight different simple actions, such as hitting the bottom of a bowl, to five chimpanzees. These researchers found that in only 5 percent of cases did the chimps imitate correctly on their first attempt. Though there are hints of comprehension in other animals, such as apes, dolphins, and birds, nowhere do we find seeing and knowing as intimately connected as they are in our own species. Do animals sometimes understand the implications of what they see? Of course they do. They run away from dangerous predators, pounce on succulent prey. But even our closest relatives do not comprehend what they see others doing in the same ways that we do.

Tests of how much a chimpanzee might comprehend of the motivations of others—tests, that is to say, of theory of mind—show marked differences between humans and nonhumans. In theory-of-mind tests, apes just don't get it. They may gradually pick up the implications of the bizarre actions of human trainers in the complicated guesser/knower and cooperative trainer/competitive trainer experiments, but they fail to show the rapid comprehension of the situation that a theory of mind ought to supply. In the simpler task of begging for food, the chimpanzee is astonishingly blind to the implications of what people can and can't see. Though the interaction between a dominant and subordinate animal in competing for food may be encouraging for those who want to believe that the chimpanzee has a theory of mind, I think the easier explanation for the subordinate chimp's performance in that situation is that it is fearful of looking at whatever the dominant animal is looking at—which does not imply that it knows why it has that fear.

We have always expected magic of mirrors. Shakespeare loved mirrors, particularly as revealing the truth about people, and referred to them often in his sonnets:

> For to no other pass my verses tend
> Than of your graces and your gifts to tell;
> And more, much more, than in my verse can sit
> Your own glass shows you when you look in it.

But mirrors also stand for distortion and confusion. Alice's looking-glass world is full of perplexity. Even after she realizes that things are spelled backward, she doesn't find it easy to make sense of "Jabberwocky." ("SOMEBODY killed SOMETHING: that's clear, at any rate" was Alice's response after reading the poem.)

Showing a mirror to another ape is a form of making animals into mirror images of ourselves. Is this Shakespeare's mirror of truth, or Lewis Carroll's looking glass of confusion? Why should we think that, even if an animal reacts to a mirror as we do, it is thinking the same thoughts as it does so? Is it really through viewing ourselves in the mirror that we know ourselves? I think not. As more species are found that react as we would to their reflection in a mirror, I think the magic will start to wear off the silvered glass. Though many species imitate each other, and a few have some understanding of what they see, for no other species is there the same deep connection between seeing and knowing that we take for granted in ourselves and in each other.

FURTHER READING

The Ape and the Sushi Master: Cultural Reflections of a Primatologist, by Frans de Waal (Basic Books, 2001). De Waal's starting points may be very different from mine, but his book is nonetheless a very stimulating and, in parts, highly informative discussion of culture in humans and other primates.

Folk Physics for Apes: The Chimpanzees' Theory of How the World Works, by Daniel J. Povinelli (Oxford University Press, 2000). Povinelli's account of experiments looking for a theory of mind in chimpanzees has the engagement and excitement you'd expect of one who has been personally involved in this work.

Dolphins Divine

Diviner than the dolphin is nothing yet created; for indeed they were aforetime men and lived in cities along with mortals, but by the devising of Dionysos they exchanged the land for the sea and put on the form of fishes.
—Oppian, Halieutica, A.D. 200

Seeing a dolphin in the wild is like bumping into a celebrity in the street. There's a sense of excited recognition, and of one's own unworthiness in comparison to these exceptional individuals. One of the things that made living in Perth, Western Australia (where I started this book) very special was the presence of dolphins in the Swan River. Perth, population 1.4 million, is one of the few major cities in the world that can boast dolphins right in the center of town. I wouldn't want to exaggerate how often they might be spotted, but I couldn't overstate what a pleasure it was, during a lunch break or a midmorning stroll, to see a pod of dolphins bouncing in and out of the water in their characteristic fashion. Not a rarity, and yet so beautiful, so uplifting, that strangers on the footpath point them out to each other, and stop and talk for a moment, before sinking back into the anonymity of a big city.

These dolphins, and most of the others I shall be talking about in this chapter, are bottlenose dolphins, *Tursiops truncatus*. This is only one of the more than thirty species of dolphin and porpoise—never mind those other aquatic mammals, the whales—but the bottlenosed are by far the best-studied species. Dolphins and porpoises are distinguished from whales by having neat rows of teeth: whales feed by filtering water through a structure called baleen or whalebone, from which Victorian ladies' corsets were made.

But the dolphins of the Swan River are too irregular in their habits to satisfy tourists who don't have a lot of time to hang around on the water's edge. So buses ferry visitors six hundred miles north to a bay where dolphins put in regular daily appearances. The bus trip to the beach at Monkey Mia is almost twelve hours each way, and yet people are willing to suffer twenty-four hours on a bus just to see wild dolphins close up for a couple of hours, so great is the allure of these animals. Tourist literature is keen to present the dolphins of Monkey Mia as visiting the bay out of their own curiosity and eagerness to interact with visitors. The fish handed out by National Parks officers to encourage the dolphins' appearance are quietly ignored.

Why are we so excited about dolphins? It's hard to look at the big grin on a dolphin's face without smiling back. It's hard to see a dolphin leap out of the water without remembering the joys of splashing in water as a child. Dolphins have so many natural advantages. If dolphins were people, we know they would be happy, carefree children dwelling in a perpetual springtime. And it's hard not to assume that animals are the way we think we would be if we were they.

The truth is, dolphins are nothing like us. Sure, they are mammals, as we are, but then so are aardvarks, platypuses, and camels (actually, camels are among the dolphin's closest relatives on land). Dolphins are fascinating precisely because their world, and their means of comprehending that world, is so different from our own. Dolphin popularizers have missed the point: dolphins are different, and it's the differences that make them so interesting.

Some 50 million years ago a few mammals went back into the water from which all animals had originally come. This small step had radical ramifications for every aspect of these protodolphins. The body of the mammal had to be completely reshaped for aquatic life, but so too did the mind. Far from being the happy children prancing in the waves that we imagine we see, the life of dolphins is actually at least as weird as that of the bats we met in chapter 6.

FIRST, BUILD YOUR DOLPHIN

How did dolphins get to be the way they are? Every aspect of the standard mammal needs modification for survival in the ocean. Land mammals are not hydrodynamic and can only swim slowly. Furthermore, many aspects of mammal physiology need substantial modification for success at sea. Mammals breathe air and drink fresh water: how is this to be achieved in the ocean? Most mammals rely predominantly on smell and vision to find their way around (though hearing is important too). Both smell and vision are of very limited effectiveness underwater. The aquatic world is altogether a very alien one for a mammal—how do dolphins survive and thrive there? What can we know about the life of the dolphin in this strange environment?

Aristotle recognized that dolphins were not fish. In his *History of Animals* (350 B.C.), he identified that dolphins have proper bones, not fish spines (we now know that dolphin flippers contain the bones for five fingers). They give birth to live young that suckle on their mother's milk. He noted that they cannot breathe underwater but have lungs like land animals and must return to the surface to gasp fresh air. Their breath is warm and moist like that of mammals, not cold like fish. The eye of a dolphin is also lively and somehow warm and familiar, quite unlike the cold eye of a fish. Finally, dolphins, again unlike fish, have a small tongue, somewhat like a pig's. Perhaps it is this tongue that has given us the other name sometimes used for dolphins: porpoise. "Porpoise" is derived from the Latin *piscis porcus,* meaning pig fish. A most unflattering name, but one not entirely off the point when we come to consider the dolphin's evolution. Properly speaking, a porpoise is not a dolphin but a closely related animal (smaller, with a head not shaped into a beak and with flatter teeth), but formerly these distinctions were not clearly made and the two names, dolphin and porpoise, served as synonyms.

When the ancestors of dolphins returned to the water, bats were just appearing on land, and monkeys and apes had not even been thought of. It is believed that dolphins evolved from mesonychids, somewhat doglike animals that roamed around riverbanks. For reasons now lost in the mists of prehistory, these animals began to spend more and more time in water—perhaps to find food, perhaps to escape predators, perhaps both or for some other reason.

The earliest known dolphin fossils are of a beast known as *Pakicetus* from around 52 million years ago. *Pakicetus* was clearly an aquatic mammal but, unlike later dolphins, it still had hind legs, and its skull did not show the flattened and telescopic shape of the modern dolphin. True modern dolphins start to show up in the fossil record around 36 million years ago. These are called *odontocetes*, because of their teeth (Greek, *odont*). By coincidence, modern dolphins have been around for about 5 million years—our best estimate for the age of the human ape.

Antony Alpers, a New Zealander inspired to write a delightful children's book on dolphins by a wild dolphin he knew of that was happy to play with children, offers the following suggestion for how to visualize the evolution of a land-based mammal into a dolphin. It is so attractive that I will quote it at length, first reiterating Alpers's warning that this is "not what happened in nature but only the appearance of what happened":

> Take a lump of plasticine about twice as big as your thumb, and make a rough four-legged animal, not too much like any that you know. . . .
>
> Now imagine that this animal for some reason is going to start seeking its food (and maybe refuge from some enemy or competitor) in the water, starting possibly in swamps and rivers or possibly by the sea. Begin a sort of smoothing-out process, working backward from the head, as if your fingers were imitating the flow of water over the animal's body . . .
>
> Starting at the front, push in the head so that the neck fills up, and wipe the ears away, leaving only pinholes. Flatten the front legs into

> flippers, and take the hind legs off entirely. *Keep them near you.* Now attend to the tail region. First, pinch it from the sides so that it is narrow like the hinder part of a fish. . . . Then pinch the *end* of it in the other direction to give the beginnings of horizontal tail flukes. This is where you will need the discarded plasticine; adding the extra bits, form a dolphin's tail, with its gracefully curved edges and a notch at the center. Shape the dorsal fin as well, and do it with style.
>
> Now go back to the nose. Shape it into a dolphin's beak and, with fingers as dexterous as those of the Creator, simultaneously cause the nostril holes to move to the top of the head, uniting them there in a single crescent-shaped blowhole. Add finishing touches, such as shaping the flippers to their proper form, and perhaps supplying the smile of a *Tursiops*.

Obviously Alpers's story is fanciful, but it gives some feel for the process involved in adapting to an aquatic environment. Dolphins' closest relatives on land today are the even-toed ungulates, such as pigs, deer, and hippopotami.

Having a hydrodynamic shape is only a start. For a mammal to survive in the ocean it must also breathe without drowning, obtain fresh water, and swim fast enough to catch fish.

Most people know that dolphins must surface regularly to gulp fresh air. The blowhole on the top of a dolphin's head is actually its left nostril. (The right nostril terminates under the skull in a series of sacks that are believed to be involved in sound production—of which more in a moment.) Dolphins do not have larger lungs for their body size than humans do: bigger lungs would increase buoyancy and make diving more difficult. Instead, dolphins have powerful muscles in their airways to force air through more efficiently. Even so the record for a dolphin dive is only 7.25 minutes; the record for a human is just over 6 minutes. But dolphins regularly and without any apparent discomfort dive for around 5 minutes to depths of over 650 feet. The Ama "sea-maidens" of Japan harvest shellfish from the seabed without diving gear. Even they succeed in diving repeatedly to depths of only forty feet and for just one minute at a time. As well as diving deeper and holding their breath longer without difficulty, dolphins also have adaptations in their circulation to reduce the risk of nitrogen precipitation, the cause of the excruciating condition known as the bends.

What about fresh water, essential to mammal life—how do dolphins get that? This is still not entirely clear. It is known that dolphins win water from the metabolism of glucose (glucose plus oxygen yields carbon dioxide plus water). Dolphins also have very large kidneys, which look like bunches of grapes. Each "grape" is actually a minikidney of its own. Their kidneys are so effective that they may be able to extract salt from seawater. Dolphins also get "drinking water" by extracting water from some of the less salty fish in their diet.

Though dolphins may not need to drink fresh water, they also don't want to swallow more seawater than can be helped. To prevent this, they possess a special structure in their larynx. The larynx in humans is the site of the vocal chords. In dolphins the larynx holds a special adaptation called the "goosebeak" (because of its shape). This is a stopper that fits snugly into the nasal passage and serves two functions: it separates breathing thoroughly from swallowing, so dolphins don't choke when they swallow fish whole, and it enables the dolphin to make sounds while breathing freely at the same time.

Maintaining body temperature, as mammals must, is also more of a problem in water than on dry land. Try this experiment. Take all your clothes off and sit in a room at 70°F. You can try this for an hour, but you'll probably feel comfortable doing it indefinitely. Now fill a bath with water at the same temperature and immerse yourself as completely as possible. That's nothing like as comfortable, right? Water has a much higher capacity for heat than air does, so the warmth of your body gets taken away much faster. Consequently aquatic animals need special adaptations to keep their core warm.

In this regard the first thing dolphins (like whales) have going for them is their size. Bulk helps keep the heat in. Dolphins are pretty big mammals (around 10 feet long and over 240 pounds in weight); dolphins smaller than this are only found in warmer water. Then they have a thick layer of fat (blubber) that acts as an insulator under the skin. Only the fin (the one on top), flippers (one on each side toward the front), and flukes (tail stabilizers) are uninsulated, and the possibility of heat loss in those extremities is reduced by special systems of blood flow that ensure that blood on its way out to the extremities is already cold before it gets there.

Swimming is very hard work—for people. There is barely a pedestrian in the world who could not outrun the world's fastest human in the water. Alexandr Popov holds the world record for the fifty-meter freestyle, with a speed of just 5.2 miles per hour. Dolphins, with their highly hydrodynamic bodies typically cruise at about 4 to 7 miles per hour. The world speed record for a dolphin is 18.5 miles per hour.

Though dolphins can put on a burst of speed in the water when they want to, they avoid this effort if possible. As sailors well know, dolphins love to surf the bow waves of ships and save themselves a lot of effort that way.

Why do dolphins bounce in and out of the water? In order to breathe dolphins must come to the surface. But the surface of the ocean is a bad place to try and swim because of the constant buffeting from surface waves. Experiments with dolphin-shaped and -sized dummies dragged behind boats have demonstrated that, if a dolphin has to come up to the surface, it is more efficient to jump completely clear than to bob along in the waves.

So dolphin bodies have departed substantially from the standard mammalian plan. Their shape and structure is radically different, and their physiology shows many changes too. All this is interesting, but so far we have just been talking natural history. What about the mind of the dolphin? Does an alien mind come with this strange body? As with bats, we may not be able to know what it is like to be a dolphin, but we can and do know a good deal about how dolphins live and how they perceive their world. And we know that the demands of living in water have led to the development of radically different senses from the ones that land mammals depend on.

Dolphin vision is not great. Light underwater is dim. Divers consider forty feet of visibility pretty good. So although dolphins can see, vision isn't their dominant sense. Taste and smell are of no use in water. So what is a mammal to do? Listen. Sound in water travels at four and a half times its speed in air—and it also travels much further. An underwater explosion off Hawaii can be heard forty minutes later by hydrophones in San Francisco.

The first person to suggest that hearing might be important to dolphins was Arthur McBride, the original curator of the world's first dolphinarium, Marineland, in northern Florida. Marineland was constructed in 1938, originally as Marine Studios, a film studio specialized for shooting underwater movies. Basins were built with large windows in their sides offering an exceptional view of dolphins underwater. Pretty soon the public starting taking an interest in this superb close-up view of these wonderful animals. As well as perfecting systems for pumping fresh seawater through the tanks and taking care of the dolphins, McBride was instrumental in the development of methods for training dolphins to jump and twirl and do all those other tricks that visitors to dolphinariums have come to expect. In the 1940s and '50s Marineland was a major tourist destination.

Marineland still exists. But only just. By coincidence I now live less than a two hours' drive from the site. Marineland is too modest for modern tastes. Compared to the theme parks just down the road at Orlando, Marineland seems small beer. What a tragedy. With hints of Floridian art deco in its buildings and a superb site right on the beach of an Atlantic barrier island, it would be a terrible shame if Marineland went extinct. Not to mention its stellar role in the history of our understanding of aquatic mammals: It was at Marineland that dolphin mating and birth were first observed. The suckling of a dolphin calf by its mother, as well as a mother holding a dead infant above water; the importance of hearing to dolphins; their social interactions; the possibility that they communicated through characteristic whistles—all of these observations were first made at Marineland. In 1940, when he had had dolphins in the tanks less than two years, McBride published a delightful article in *Natural History* in which he reports all these startling new discoveries.

A LITTLE ECHO

In these early years Marineland encouraged scientists to come down and examine their dolphins. Winthrop N. Kellogg, who, as

Marineland of Florida, circa 1939. The scene of so many wonderful discoveries about dolphin behavior, it has hardly changed these sixty-odd years.

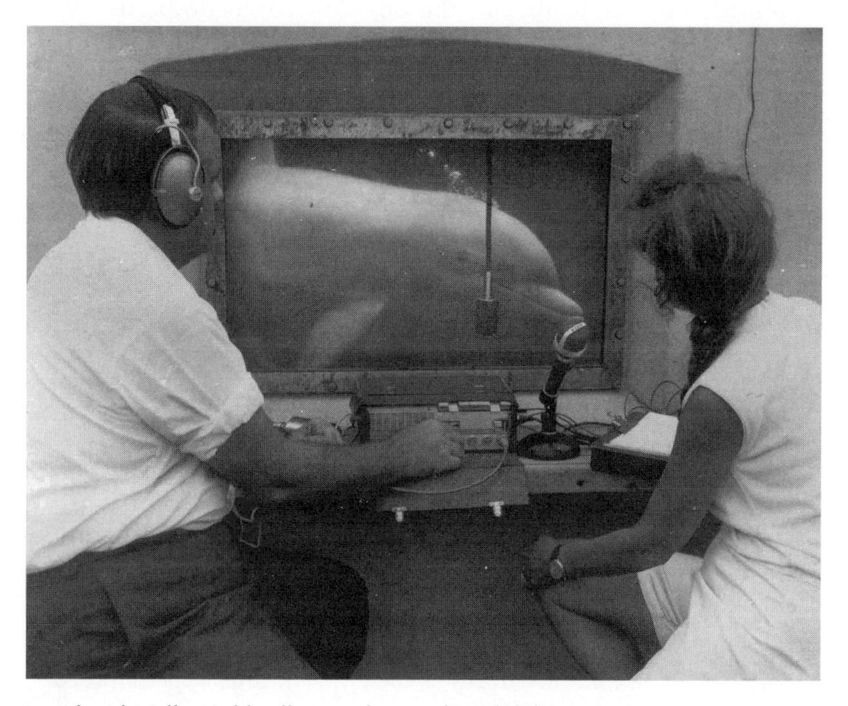

David and Melba Caldwell at work recording dolphin sounds at Marineland. (Marineland of Florida)

we saw in chapter 5, had adopted a chimpanzee infant in 1929 from the Orange Park primate center just outside Jacksonville, Florida (less than an hour's drive up the road from Marineland), was one of the first to arrive. By 1950 Kellogg was a professor at Florida State University in Tallahassee and conveniently situated to study the captive dolphins at Marineland as well as free-living dolphins in the Gulf of Mexico. Kellogg picked up on McBride's suggestion that dolphins might have echolocating abilities. Observing how bottlenose dolphins caught in a net in brackish water at night would succeed in finding the part of the net where the floats had been pulled beneath the surface of the water and created a passage to freedom, McBride noted that, "[this] behavior calls to mind the sonic sending and receiving apparatus which enables the bat to avoid obstacles in the dark."

McBride died in 1950, but the possibility that dolphins might use sounds as bats do to echolocate was picked up by his successor at Marineland, Forrest Wood, and by Kellogg. The U.S. Navy made special underwater microphones—hydrophones—available to Kellogg and Wood. Wood was the first to report underwater ultrasonic vocalizations by the captive dolphins at Marineland. Kellogg described the excitement of hearing the underwater sounds of wild dolphins for the first time while out on a boat fitted with a navy hydrophone:

> The intermittent tapping or sputtering which had been barely discernible from the speaker when the animals first turned in our direction grew in intensity and in continuity as they approached. When emitted by a single porpoise alone [Kellogg always referred to dolphins as porpoises], this noise—as we had learned before—is a concatenation of clicks or clacks such as might be produced by a rusty hinge if it were opened slowly. It was soon apparent, however, that a number of the animals were making the sounds together, and more seemed to join the chorus as they came nearer. Superimposed upon this increasing clatter was an occasional birdlike whistle resembling the "cheep" of a canary.
>
> As they came still closer, the sputtering noises continued to grow louder and still louder. Taken together, they suggested the roar of an

> approaching railroad train, except perhaps that they were more irregular. By the time the group was about ready to make its final dive, the crescendo from the speaker in our boat had become a clattering din which almost drowned out the human voice.

Through the 1950s Kellogg arranged for the construction of a special dolphin tank, intentionally filled with very murky water, and carried out a number of tests showing that the dolphins could identify objects and navigate around obstacles in the brackish water, even on pitch-black nights. Kellogg and his coworkers demonstrated that under these conditions dolphins could distinguish a desirable fish (spot) from an undesirable one (mullet) that differed in size by about six inches and discriminate objects of the same approximate size but made of different materials (a spot from a human hand). Kellogg and colleagues were also able to show that dolphins, in dirty water and complete darkness, could swim around a set of poles placed into the water approximately eight feet apart.

Throughout these and several other studies, Kellogg recorded the underwater vocalizations of the dolphins using hydrophones. He found that, just as had been observed in studies of echolocating bats, the dolphins would never make a move for an object without first issuing a series of sonar pulses, and the "pinging sound-trains" became more frequent as the animal approached the object.

One of the interesting things about dolphin echolocation, compared to the same skill in bats, is that nobody ever guessed that they had it. Nothing in ancient Greek myth suggested that there were magical perceptual abilities in the dolphin. Though the ancients knew that dolphins made sounds, the human ear cannot easily hear sounds under the surface of the ocean, and so underwater sounds were largely ignored until suddenly becoming of military importance during World War II. It was at this time that a method of detecting enemy ships by listening for underwater engine noise came into being. In the winter of 1942 hydrophones were installed at Fort Monroe, at the entrance to Chesapeake Bay on the East Coast of the United States. On the whole the system worked well at detecting the approach of ships, but in the following spring the hydrophones suddenly started picking up a noise like a chorus of pneumatic drills—no ship was in sight. By evening the noise was louder than that of any ship in the bay. It transpired that the hammering was coming from schools of croakers returning to the bay after spawning out at sea. The issue of natural underwater noise instantly became of paramount military importance, because by then acoustic mines and torpedoes had been developed that were supposed to be detonated by enemy ships: the very real possibility arose that these might be set off by fish or dolphins. Surveys were carried out cataloging the noises of different aquatic species (this knowledge has also proven useful in peacetime, to fishing fleets looking for fish).

War being a great stimulus to creativity, it was around the same time that the first human sonar systems were developed. "Sonar" stands for *so*und *n*avigation *a*nd *r*anging and was developed for submarines because visibility underwater is so poor. Furthermore, the then newly developed system for detecting direction through radio signals—radar (*ra*dio *d*irection *a*nd *r*anging)—will not work underwater.

The principle of echolocation (or sonar) was introduced in chapter 4. Briefly, a source emits pulses of sound, and a detector, usually attached to the source, listens for the echoes. These echoes are changed in amplitude (loudness) depending on the distance from the source of the object they have been reflected from, but they are also changed in frequency (pitch) depending on the movement of the object relative to the transmitter and receiver.

Subsequent research has found that the "creaking-door," "rusty-hinge," rapid-clicking sounds that Wood and Kellogg described are actually very rapid series of sound pulses. Each burst of sound lasts less than a tenth of a millisecond (0.0001 sec). But each pulse, although very short, contains a wide range of frequencies of sound energy. The lowest frequencies are just at the upper limit of what the human ear can hear (around 20 kHz), but the peak energy in the click is above 100 kHz—three octaves beyond the limits of human hearing. The interval between clicks is varied to allow time for the dolphin to hear each echo before emitting the next click. Whitlow Au of the Naval Ocean Systems Center on Kailua

Hawaii has found that the interval between clicks emitted by dolphins increases fairly linearly from about 50 msec for an object sixty-five feet away, to 200 msec for an object four hundred feet away. In each case these values are just slightly longer than the time it takes the sound of each click to reach the object and return.

Kellogg and his coworkers suspected correctly that the dolphin's clicks were generated beneath its blowhole. The blowhole on top of a dolphin's head is nothing but a modified nose. The nostril has moved back from the front of the head to its top, where it is much more effective in protecting against drowning, and has developed a special flap to prevent the entry of water. Although dolphins can be trained in captivity to make noises (like a Bronx cheer or raspberry) by blowing air over this flap, this is not how the sonar clicks are generated. Though there is as yet no definitive consensus on the exact source of the dolphin's ultrasonic vocalizations, it is understood that the source of the clicks is in the nasal tract, a few inches beneath the blowhole.

Although sound travels extremely well underwater, and using high-pitched ultrasound gives dolphins the advantages of accurate direction finding that we considered in chapter 4 when discussing bat echolocation, dolphins are confronted with a problem that bats don't need to face: the much greater resistance of water to sound energy.

Next time you are underwater with a friend, try making noises at each other. If your friend blows a raspberry at you while you are both underwater, you will not hear much, even if she is very close. You will see bubbles come out of her mouth but hear no sound. Now this is partly because your ears are not adapted to hearing underwater, but it is also because it is very difficult to get sound energy from the relatively conductive medium of the air in your mouth into the much denser medium of water. This is also the reason why, if you shout from outside a swimming pool at somebody whose ears are underwater, they will not hear you—most of the energy in your shout is reflected back up at you from the water's surface.

The quality of a substance that determines how readily sound enters it is called its acoustic impedance. The problem of getting a sound generated in air (low acoustic impedance) into water (high acoustic impedance) is called impedance matching. Although we can hear dolphin sounds coming out of the animal's blowhole, this is not how the sounds are emitted under water. Under the water's surface a sound emitted through the dolphin's blowhole would be as effective as your friend blowing a raspberry: it would not carry into the water.

The area under a dolphin's forehead and in front of its blowhole is filled by an object called a melon. The melon is an impedance-matching device. It contains two densities of fat. The back portion (about one-third of the total) connects to the nasal tract just by the area where it is believed that the ultrasonic clicks originate. This back portion is filled with fat of a lower density. The front two-thirds of the melon is filled with higher-density fat. This high-density fat has approximately the same acoustic impedance as seawater. Thus, although we can hear dolphin clicks coming out of the blowhole, the majority of the sound energy is transmitted from the nasal tract through the melon directly into the water in front of the dolphin. As well as matching the impedance of seawater so that the sound is transmitted efficiently, the melon also acts to focus the sound energy into a narrow, directed beam (though claims that this beam can be so powerful as to kill fish have never been supported).

But efficiently emitting sound underwater is not enough. For sonar to work there must also be an effective receiver. Try getting your underwater friend to bang two pieces of metal or hard wood together while you are both underwater: You still don't hear much do you? There is no problem here of impedance matching in the transmitter (the wood or metal is about as dense as the water), but the problem is still impedance matching, this time in the receiver. You may be able to improve matters by swallowing, but if you move deeper, or come back up to the surface, you are going to need to keep swallowing with every change in depth. If you bounced in and out of the water like a dolphin this would be an insurmountable problem. The standard ear of the land mammal is not set up for efficiently receiving sound if the outside pressure is continuously varied.

Dolphins have no external ears, because these would create drag while moving through water. But they also have almost no ear canal (a vestigial pinhole channel remains). It seems most likely that dolphins pick up their sonar echoes not through their pinhole ear channels but through their lower jaws, though this question, like the question of where dolphins produce their ultrasounds, is still debated. The lower jaw of the dolphin is connected to another package of the same high-density fat with excellent sound-transmitting qualities that is found in the melon. This fatty channel in turn connects to the middle and inner ears. These structures are kept in auditory isolation from the rest of the head in bone cases held separated from the skull by cartilage and connective tissue. This isolation of the ears from the rest of the skull probably enhances the stereo separation of sounds which is important in localizing echoes. Stereo separation is a more difficult problem underwater because the speed of sound through water is much higher than through air.

To test the hypothesis that dolphins hear through their jaws, Randall Brill and his colleagues, working out of the U.S. Space and Naval Warfare Systems Center in San Diego, developed a special kind of earmuff—an earmuff for a jaw. First, the dolphins were blindfolded and trained to find objects underwater using only their sonar. Then, on different tests one of two different masks was fitted around the dolphin's jaw. One mask was made of sound-transparent synthetic rubber; the other was constructed of a different type of synthetic rubber that blocked sound. On trials with the sound-transparent rubber mask, the dolphins continued finding objects even though blindfolded. On trials with the sound-blocking rubber masks, however, their performance was greatly disrupted.

Dolphin sonar appears superficially simpler than bat sonar. Dolphins only have a single type of click, and they vary only its amplitude and the interval between clicks. Bat's echolocation calls, by contrast, can vary in frequency and duration. Nonetheless, dolphins actually outperform bats in many tests of the usefulness of sonar.

Dolphins can discriminate the thickness, size, and shape of objects made from identical materials (such as a metal disk inch one-eighth of an inch thick and another that is one-twelfth of an inch), as well as the materials of which identically sized and shaped objects are made. Dolphins were first found to discriminate copper from brass disks of identical size. Further research, particularly in the former Soviet Union, uncovered an ability to discriminate materials ranging from wax, rubber, and plastic to metals like lead, steel, and aluminum. There now seems no limit to the list of materials that dolphins can identify. Much of this research was carried out at military establishments in Hawaii and San Diego in the United States and at Black Sea ports in the former Soviet Union. The U.S. Navy still employs dolphins, and U.S. dolphins are presently on mine-sweeping duties in the Persian Gulf. With the collapse of the Soviet Union, Russian dolphins, which had been trained not just to find mines but also to plant them, found themselves in the cold waters of the free market. In March 2000 the BBC reported that ex-Soviet dolphins had been sold to Iran. Their trainer, Boris Zhurid, went with them but refused to tell the BBC reporter what use, if any, the dolphins would be put to. "I am prepared to go to Allah, or even to the devil, as long as my animals will be OK there," was all he said. There would seem to be a real possibility now of U.S. and former-Soviet dolphins settling Cold War scores in the Persian Gulf.

Bats do not possess this ability to identify the materials out of which objects are made. This is because sound in air is reflected back from denser materials. But the sounds dolphins create underwater are already transmitted effectively (through the melon) into a high-impedance medium, seawater. Consequently, when these sounds reach solid objects, they are not reflected back entirely. Part of the sound energy penetrates some way inside the object before being reflected and thereby reveals some of the inner structure, just as medics use ultrasound to image fetuses in a mother's womb. Not that I am suggesting that dolphins "see" a picture in that kind of detail (nor is there any support for stories that dolphins can tell that a woman in water is pregnant), but the quality of echo they experience is clearly changed in some way depending not just on an object's size and shape but also the material of which it is made.

Unlike the moths that echolocating bats are interested in, which, as we saw in chapter 4, use blocking sounds to try and confuse the foraging bat, the fish that dolphins feed on do not appear to have any strategies to interfere with dolphin sonar. On the contrary, plenty of fish species are so noisy in the water that the dolphins don't even need their sonar; they just home in on the noises the fish make. Like journalists listening in for excitement on police radio frequencies, dolphins channel-surf through the sound frequencies of fish.

IF I ONLY HAD A BRAIN

It was the massive brain of the dolphin that first caught John Lilly's attention. Lilly was one of the many young scientists excited by what he saw at Marineland. Originally a brain scientist before discovering isolation tanks and LSD in the 1950s and '60s, Lilly became, to quote his web page, "an unparalleled scientific visionary and explorer, . . . [who] made significant contributions of psychology, brain research, computer theory, medicine, ethics, delphinology, and interspecies communication."

Lilly was convinced that dolphins communicate with each other via a sophisticated language based on their sonar vocalizations. In a 1962 book Lilly predicted that we would have decoded this language and be able to converse with dolphins in two-way communication within twenty years. In 1983 he told an interviewer with the popular science magazine *Omni* that the project would need another five years. Lilly died in 2001, his ambition of talking with dolphins still unfulfilled. Most of the stories one hears about dolphins in the popular press—that they have superhuman intelligence, communicate at an advanced level, engage in sex just for recreation (whatever that means)—originated with Lilly.

The tragedy of Lilly is that he started out as a serious investigator of dolphin brains and behavior. Readers of his entertaining first (1962) book on dolphins will be struck by his frank fascination with these animals (and with the number he succeeds in killing while trying to record their brain waves). Lilly was present at the construction of the first dolphinariums and tells the story well. He was the first to notice that dolphins sleep with only half of their brain at a time—a discovery confirmed by subsequent investigators. But somehow he became a victim of his times, and by the time he published another book on dolphins (1978), something has gone badly wrong with Lilly.

On the first page he states that dolphins and whales "with [their] huge brains are more intelligent than any man or woman." He proceeds to catalog the many evidences that dolphins are smarter than people. Dolphins do not draft each other to fight wars against other groups. They do not capture other species for their own entertainment. They are "sensitive, compassionate, ethical, philosophical, and have ancient 'vocal' histories." Lilly believed that if he played back dolphin vocalizations at half or quarter normal speed, he could make out words that they were saying to him. He acted as consultant on the movie *Flipper* and the TV series of the same name. He believed that once human-dolphin communication was established, cetaceans might be represented at the United Nations; the U.S. government would set up a Cetacean Communication Department; the Library of Congress, "a new division given to recording the history, philosophy, ethics, and science of the cetaceans."

Lilly is easy to ridicule (in 1967 he even tried giving LSD to dolphins), but his influence was not all bad. Lilly's books and articles encourage people to take an interest in the potential of dolphins, and his activism was doubtless helpful in raising public awareness of the damaging impacts of the whaling industry and the effects of tuna nets on dolphins. But the problem is that Lilly's mythologizing has gained wider distribution than the facts that are known about dolphin behavior. That repository of contemporary myths, the World Wide Web, contains many more sites about dolphins that take Lilly's conjectures as fact than it has references to what is really known about these animals. The dolphin-hugging business is a major industry. There are numerous organizations committed to the miraculous cures that contact with a dolphin can produce in handicapped children.

Lilly and his acolytes like to emphasize the size of the dolphin's brain. Having a brain makes you brainy—smart. Our brain is our most cherished organ. It has given us intelligence, reason, morality, and culture. No other organ in the human body compares impressively in size to the corresponding organ in other animals; our brains are unique. At around three pounds, we have, for our body weight, about three times the brain of a typical primate. Dolphins have even larger brains (around three and a half pounds), though, when corrected for body weight they don't quite match us in brain size (dolphins weigh on average twice as much as human beings).

But not all parts of our brain are equally impressive. In human evolution it is the neocortex, the folds of tissue on top of the basic brain, that have taken on truly exceptional dimensions. In most mammals the neocortex takes up between 10 and 30 percent of the total, reaching values above 50 percent only for primates, but the human neocortex takes up a magnificent 80 percent of total brain volume. Our neocortex is so large it has to be folded to fit inside the skull. If you removed your brain from your skull (don't try this at home), you would find that, unfolded, your neocortex occupies more than twice the area on the kitchen table as it does inside your head. But on this measure of human distinction, dolphins, it is often pointed out, outstrip us. The unfolded neocortex of a dolphin takes up *five times* as much space as it does packed into the dolphin skull. Though this fact appears in most popular texts on dolphins, one rather important detail is always missing. Though all land mammals have approximately the same density of nerve cells in their neocortexes, sea mammals do not. Mammals in the ocean have only about one-quarter the number of nerve cells per square inch of cortex as do land mammals. Thus, in terms of numbers of nerve cells, the functioning units, the dolphin neocortex is not so large after all.

Brains are not supernatural objects but are subject to the laws of physics like any other physical entities. As land mammals we have to carry our brains about with us all day, every day (if that seems like a funny way of putting it, remember that human infants are unable to hold up the weight of their own heads for the first two or three months of life). If we could fly, the problem of brain weight would be even more acute. But sea mammals do not have to worry about weight. Consequently, without the check of gravity, it seems that evolution has allowed the weight and volume of dolphin brains to blow out. That is not to say that the dolphin brain is not still large on anyone's terms. Just that there is nothing mysterious about the size of a dolphin's brain.

A brain is not just subject to the laws of physics but to the laws of biology too. Breathing in the dolphin is known to be under the control of the cortex, rather than the deeper brain centers used in other mammals. The dolphin's sonar system must be controlled from somewhere, and the cortex is a likely candidate for that function too. One of the brain centers that controls hearing is over 250 times larger in the dolphin brain than the equivalent structure in the human brain. Thus, the brain of the dolphin must be understood as a component of the body it inhabits and the things that body can do.

GIVE A LITTLE WHISTLE

As well as suspecting that they were using sounds to locate things, McBride also speculated that his dolphins were using sound to communicate. Husband and wife team David and Melba Caldwell spent many years at Marineland pursuing this possibility, and it is to them that the credit goes for naming and identifying the function of the dolphin's characteristic whistles. These whistles are a pleasant warbling sound, somewhat like the cheep of a canary and do not extend much beyond the range of human hearing. Each dolphin has a characteristic whistle, and the Caldwells therefore named them "signature" whistles. Unfortunately, Lilly's spurious claims that dolphins have a communication system as complex as human language have been far more widely distributed than the Caldwells' careful observation of what dolphin whistles are really for. At Lilly's feet must be placed some of the blame for the fact that when the Caldwells published their discovery that dolphin whistles are just signature calls that identify individuals and not a full-fledged "language," they were received with an "unbridled anger" that shocked them.

A young dolphin develops its own signature whistle in the first months of life, and it generally remains unchanged throughout adulthood. Interestingly, although female dolphins usually develop signature whistles quite different from those of their mothers, about half of male dolphins' signature whistles are very close to their mothers' whistles. This is probably because in dolphin society, like that of many other mammals, the males leave the mother after weaning to form all-male groups, while the daughters stick together in groups of females and young. Consequently, daughters need whistles that can be distinguished from those of their nearby mothers; sons don't.

Dolphins do not just whistle their own signature whistles; under certain circumstances they may copy other's whistles too, usually the whistles of close social contacts. Estimates of how often this happens vary greatly. Some researchers are unable to find any emulation of signature whistles, while others find that about half of all the whistles they record have been copied from another dolphin.

For a long time there was no clear consensus as to the function of dolphins' signature whistles. Could they possibly be the basis of a subtle communication system? Recent studies do indicate that the signature whistle forms the basis of a communication system, but a fairly limited one.

Vincent Janik and Peter Slater from the University of St. Andrew's in Scotland measured the emission of signature whistles by captive dolphins at Duisburg Zoo in Germany. These dolphins could be moved away from a social group and back into the group. Janik and Slater found that, as they were moved out of visual contact, the dolphins would start to emit many more of their signature whistles. It seems, therefore, that the signature whistle is a bit like calling out "I'm over here" in a darkened room. The characteristic sound of your voice tells others of your social group where you are when vision is ineffective.

But what about the practice of imitating other's signature whistles? Surely that would only confuse other members of the group? The function of signature whistle imitation is still not well understood, but one suggestion comes from an interesting study by Peter Tyack working out of Woods Hole Oceanographic Institute in Massachusetts. Tyack and his colleagues captured a wild dolphin and held it for an hour with a hydrophone close by. During the first thirty minutes, the dolphin whistled its signature whistle 520 times. During this period there were no recognizable imitations of another dolphin's signature whistle. In the second half-hour, the captive dolphin suddenly started producing imitations of the signature whistle of another dolphin in its social group. Its own signature whistles dropped slightly, to 472 in thirty minutes, but forty-seven imitations were recorded—mainly of the oldest dolphin in that group. Tyack suggests that these imitations of the oldest dolphin's whistles were the captive dolphin's attempt to call for assistance. This is an interesting hypothesis, but more research is needed before we can be sure why dolphins sometimes imitate each other's signature whistles. As well as imitating others' signature whistles, both captive and wild dolphins have been shown to recognize different individuals' signature whistles when these were played back from a tape recorder.

So Lilly's claims that dolphins have a communicative system as rich and complex as human language receive absolutely no support from any impartial study. The ability of dolphins to communicate may be little greater than the characteristic self-identifying bark of a dog.

SEX AND THE SINGLE DOLPHIN

Another strange claim of Lilly's is that dolphins are the only animals besides humans to sometimes engage in sex solely for recreation, not procreation. (Does this claim imply that all other species know that sex leads to pregnancy and offspring?) In any case, the facts about dolphin sexuality are far less cheery than Lilly claimed.

It is not only tourists desperate for a glimpse of a dolphin who take the bus to Monkey Mia in the north of Western Australia. A research team from the University of Massachusetts in Dartmouth and Georgetown University in Washington, D.C., has also been making extended visits to Monkey Mia since 1984. Over the years this group, led by Richard Connor, has built up a detailed picture of the social and sexual lives of these wild dolphins.

A dolphin calf at Monkey Mia usually stays with its mother

from three to five years. Then, if the young dolphin is male, he will go off and join an all-male group that will hunt for fish and female dolphins. If the young dolphin is female, her fate is somewhat different. Dolphin females spend less time associating with other adult dolphins than do the males; they spend most of their adult lives associating with juveniles (sound familiar?). Between five and seven years of age her menstrual cycles will commence, but she will not give birth until she is at least ten years old.

Connor and his colleagues use the term "consortship" to describe the association of a group of males with a single female for sexual purposes. This decorous nomenclature is like a fig leaf laid over the facts of sex and the single dolphin. Connor's group points out that these "consortships" were typically violent and coercive unions. Connor points out that "[m]ales used vocal threats, physical threats, and attacks to keep a female close." Female dolphins at Monkey Mia, once captured, were typically held by a male group for between two days and a week, though, on occasion, capture lasting up to twenty-eight days was recorded. Sometimes a female dolphin captured by one male group and held for a few days would then be stolen by another group of males and held by them for several days. Out of a survey of 255 consortships observed over three years, Conner and his group estimated that in at least 82 percent of cases the males used coercion to capture the female. In most of the remaining cases, coercion may have taken place underwater, where the researchers could not observe it. During the period of consortship, the female dolphin usually has sex repeatedly with all the males in the group that have captured her.

Connor's group did note that sometimes male and female dolphins would come together in a more gentle and, to human eyes, affectionate manner. But these "affiliative" interactions were the exception rather than the rule, as far as the researchers could see.

GIVE US NEW MYTHS

Dolphin sex: not recreation but coercion. Dolphin communication: not complex, just simple identifying tags. Dolphin brains: big, but not so astoundingly big. Is there nothing wonderful about dolphins after all? Of course there is. Dolphins may not be the "hobbits of the sea," as one dolphin researcher despairingly labeled the popular view of these animals, but they are still magnificent beasts. Furthermore, one of the most improbable tales about dolphins turns out to have some truth to it.

For millennia there have been stories of dolphins helping drowning swimmers and shipwrecked sailors. Plutarch recorded: "To the dolphin alone, beyond all others, nature has granted what the best philosophers seek; friendship for no advantage. Though it has no need at all of any man, yet it is a genial friend to all and has helped many." The ancient Greeks and Romans told many stories about dolphins helping gods and men. But could or would a dolphin save a swimmer in trouble by keeping her afloat and pushing her to safety? A few reports from modern times suggest there may be some truth in the ancient stories. In 1949 *Natural History* reported the story of a mature, well-educated woman from Florida who was swept out to sea by an undertow: "[A]s I gradually lost consciousness . . . someone gave me a terrific shove, and I landed on the beach, face down, too exhausted to turn over. . . . It was several minutes before I could do so, and when I did, no one was near, but in the water about 18 feet out a porpoise was leaping around." The woman also said that a man who had been watching saw a porpoise shove her ashore.

Another Floridian, a Mrs. Yvonne M. Bliss, gives an eyewitness account of being rescued by a dolphin in 1960. Mrs. Bliss fell overboard at night and, after some time treading water, saw a form in the water that she at first thought was a shark. This "sea life," as she called it, nudged her into a better position in the waves, so that she didn't swallow so much water, and guided her to where the water was most shallow and she was able to touch the bottom.

It had crossed my mind that tales of being rescued by dolphins might be the Floridian equivalent of stories from New Mexico of abduction by aliens. But I recently found a story of a British tourist swimming off the Mediterranean coast of Israel in 1996 who was saved from a shark by a pod of dolphins. There were several eyewitnesses to this event. Many of these rescuing behaviors (especially in Mrs. Bliss's story) are very similar to the behaviors of dolphins toward a sick member of their group. Typically, for example, a dolphin mother will nudge a sick infant up to the surface, and a human being would be much the size of a juvenile dolphin. This is pure speculation (size similarity notwithstanding, wouldn't the dolphin be able to tell the difference between a human being and a member of its own species?), but perhaps we shouldn't dismiss it too soon.

Another story from the ancients turns out to be true: dolphins really do hunt cooperatively with human beings. For centuries the ancient reports of dolphins helping fishermen were dismissed as fanciful, but now there are several well-documented cases. In the town of Laguna, near the southernmost tip of Brazil, for example, groups of thirty to forty fishermen work together with from one to four dolphins in the large brackish lagoons near the town. The fishermen, each with a circular net, position themselves in a single line along the shore in about three feet of water. A dolphin just off shore submerges, then reappears in a few seconds and comes to an abrupt halt before diving away from the nets and creating a surging roll of water that washes toward the fishermen. As the dolphin turns and dives, the fishermen cast their nets and catch the fish that the dolphin has herded toward them. The water is so turbid that the fishermen cannot see anything themselves, but the success of their catch shows clearly that the dolphins perceive the fish. The fishermen have learned to predict from the energy of the dolphins' actions whether many fish are on their way or just a few. In the ensuing confusion as the fish try to escape the fishermen's nets, the dolphins themselves get plenty of fish for their efforts.

Not all dolphins help the fishermen. Dolphins that fail to assist at the catch are called "ruim": bad dolphins.

The pity of Lilly and his influence is that we don't need to suspend our critical faculties to see that dolphins are something special. The reality of what dolphins are, how they live, and what they have meant to human beings over millennia is surely fascinating enough.

As far back as written records go, people have been impressed by dolphins. "Playful," "joyful," "intelligent," "friend to mankind,"—these epithets have been given to dolphins for thousands of years. The ancient Greeks and Romans knew dolphins well (there must have been many more of them in the Mediterranean in ancient times) and told numerous stories about them. Dolphins appear on plates, coins, wall frescoes, and floor mosaics from the fifth century B.C. onward. The ancients recognized that there was something different about dolphins, something very special.

These ancient myths are beautiful stories, but scientists would normally assume them to be false. Romantic stories about magical creatures are usually much more a reflection of what we want animals to be than of what they are in reality. And for many years any suggestion that dolphins could have any positive interactions with human beings was frowned upon as romantic and anthropomorphic. But it turns out that the ancients knew a thing or two about these strange animals in the ocean. Dolphins may indeed rescue people who are drowning and certainly can help fishermen.

Though their watery lives are foreign to land mammals like us, dolphins can make sense to us in certain ways. Being ourselves social hunters, in need of comrades to succeed in the hunt, we can identify with their social lives. As swift, intelligent, and fascinating animals, they intoxicate us. But they are not like us. John Lilly's very successful attempts to convince the broader public that dolphins are floating hobbits are a deep disappointment and frustration to scientists interested in what dolphins really are.

The pleasure we take in dolphins, their playful smile and joyous gamboling in the waves, says much more about what it is to be human than what it might be like to be a dolphin. And yet dolphins really are fascinating and wonderful animals—precisely because they are so different from us, so other. Though Oppian's premise, that dolphins are descended from men, may be false, his conclusion—"Diviner than the dolphin is nothing yet created"—may be allowed to stand.

FURTHER READING

The Bottlenose Dolphin: Biology and Conservation, by John Reynolds III, Randall Wells and Samantha Eide (University Press of Florida, 2000). An excellent and up-to-date summary of what is known about dolphins from all possible angles. It is also well written.

Porpoises and Sonar, by Winthrop Kellogg (University of Chicago Press, 1961). Though now somewhat dated, Kellogg's memoir carries the excitement of the beginnings of research into dolphin sonar.

Cetacean Societies: Field Studies of Dolphins and Whales, edited by Janet Mann, Richard Connor, Peter Tyack, and Hal Whitehead (University of Chicago Press, 2000). This volume, compiled by a team of leading dolphin researchers, is a definitive survey of the current state of our understanding of the lives of dolphins and whales in the wild.

U.S. Navy Marine Mammal Program [home page]. <www.spawar.navy.mil/sandiego/technology/mammals>. This web site offers an interesting history of the military's use of dolphins and what has been learned in their research programs.

Earth Coincidence Control Officer John C. Lilly. <deoxy.org/lilly.htm>. This page on the late John Lilly contains a variety of documents relating to his theories about dolphins, including audio of a dolphin saying "Hello Margaret"!

9

Sandwiches to Go

Back on the Isle of Wight last summer, everything had returned to the tranquility I remember. Boots The Chemist has been renovated back to its original bland splendor. The farmers on the hills tend their animals: the town-dwellers get their meat at the supermarket or butcher's shop. Fishing on the piers or out in boats is still a popular hobby. I didn't notice much excitement about animal rights issues: people there have their own concerns. Barry Horne's passing in November 2001 went largely unremarked on the island. Even on the mainland, the revenge attacks threatened by Horne's supporters against scientists who use animals failed to materialize.

It is tempting to dismiss Horne as a crackpot, but it is also possible to view him as an individual exceptionally true to his convictions. Australian philosopher Raimond Gaita, in a fascinating and moving meditation on human-animal relations, *The Philosopher's Dog*, makes the point that although animal rights activists claim to believe that the slaughter of animals for meat is equivalent to murder, he has never met anyone who acted as if they really believed it. I would credit Horne's self-sacrifice as indicative that he really believed killing animals was murder.

Could killing animals really be as evil as killing people? That would have to depend on what you believe animals experience. What kind of mental lives do they have? Few scientists are willing to stand up for the general notion that all nonhuman animals share human experience and so should share in human rights, but there is a movement to protect certain species from human exploitation that has garnered some very high-profile support. This is the Great Ape Project.

The Great Ape Project was founded by another Australian philosopher (now at Princeton), Peter Singer, together with Paula Cavalieri from Milan. Signatories to their "Declaration on Great Apes," which demands legal personhood for these animals, include renowned animal scientists such as Richard Dawkins, Jared Diamond, Jane Goodall, and Francine Patterson, along with many less widely known but still highly respected scientists. According to the Great Ape Project and related groups, such as Steven Wise's Center for the Expansion of Human Rights, chimpanzees and the other great apes are so genetically similar to us that they must share all the important psychological qualities that entitle us to have rights before the law, qualities such as a capacity for language and self-awareness. They argue that great apes should be given legal personhood and some of the rights now reserved for humans. They have had some successes: Germany and Switzerland have guaranteed certain rights for animals in recent years, and the government of New Zealand also seriously entertained the idea.

The great apes are certainly in trouble: chimpanzees, bonobos, gorillas, and orangutans are all declining in numbers as their habitat is encroached upon by human activities. In addition to the problems they experience in their home countries, chimpanzees are used in medical research on hepatitis, AIDS, and several other diseases. Being a subject in an experiment on hepatitis or AIDS is certainly not pleasant. Stephen Wise in *Rattling the Cage* offers a bitter description of the last days of Jerom (to whom Wise's book is dedicated). Jerom is a chimp who died on February 13, 1996, of AIDS after having been intentionally infected with three strains of HIV. Obviously, dying of AIDS is a horrible way to go, for man or chimp.

LEGAL PERSONHOOD REVISITED

Animals were legal persons once. Once upon a medieval time, animal rights and animal responsibilities were taken very seriously indeed. In medieval Europe thoroughgoing legal proceedings were instigated against animals that damaged property or harmed people. It is hard now to understand the motivations behind these actions, but the full edifice of the law was brought to bear on insects that plagued crops, cocks that laid eggs, foxes possessed by the devil, pigs that ate children, and other misbehaving beasts.

Sometimes it seems that the main motivation for the case was to establish the defense lawyer's reputation for ingenuity—a situation not unlike celebrity trials in the United States today. Bartholomew Chassenée, a French jurist of the sixteenth century, defended some rats put on trial at Autun on the charge of having destroyed the barley crop of that province. He insisted that, since the defendants were widely dispersed around the countryside, they could not reasonably be expected to appear in court after a single summons. When the rats still didn't appear after a second summons, he excused their nonappearance by appealing to the danger they would expose themselves to if they were to leave their hiding places to try and reach the court, because so many cats, their mortal enemies, lay in wait for them in the town. Unfortunately, the outcome of this exercise in legal time wasting has not been passed down to us.

Often it was the religious authorities who took action against misbehaving beasts. In the autumn of 1487 and again in September 1488 the Grand Vicars of the Cardinal Bishop of Autun ordered their curates to make processions through every parish to warn the slugs three times to cease bothering the people by "corroding and consuming the herbs of the fields and the vines," and to depart; "and if they do not heed this our command, we excommunicate them and smite them with our anathema."

Perhaps part of the reason why so much effort was expended on determining the guilt or innocence of animals was because of a difficult religious dilemma that confronted medieval Europeans. Sometimes animals committed crimes because they were agents of the devil. In such cases they had to be excommunicated, anathematized, and done in, ceremoniously and mercilessly. But problematic animals could also be God's just punishment of mankind for its failure to live according to his laws. In such cases the only solution was penitence and repentance. For a man to punish an animal that God himself had sent to mete out just punishment would be sacrilege. So this tricky dilemma may help explain some of the energy that went into trying animals: it was very important to understand not just what an animal had done but why it had done it.

In the annals of animal trials, pigs were particularly common defendants. On June 14, 1494, a young pig was arrested for having "strangled and defaced a young child in its cradle." A witness reported that "said pig entered during the said time the said house and disfigured and ate the face and neck of the said child, which in consequence of the bites and defacements inflicted by the said pig, departed this life." The judge deliberated and pronounced sentence: "We, in detestation and horror of the said crime, and to the end that an example may be made and justice maintained, have said, judged, sentenced, pronounced and appointed, that the said porker, now detained as a prisoner and confined in the said abbey, shall be by the master of high works hanged and strangled on a gibbet of wood near and adjoinant to the gallows and high place of execution belonging to the said monks." Perhaps because of their close proximity to human habitation, there were many other cases of pigs killing children. In 1394 a pig was hanged at Mortaign, France, for having sacrilegiously eaten a consecrated wafer.

The people of these times even believed in werewolves. In 1685 a wolf preyed upon herds around Ansbach, Bavaria, apparently even devouring women and children. The locals believed the wolf was the reincarnation of a deceased burgomaster (obviously not a popular man). Once they had finally, and with great difficulty, killed the beast, they dressed its carcass in a tight suit of flesh-colored cere-cloth resembling human skin and adorned it with a chestnut brown wig and long, whitish hair. The snout was cut off, and in its place they strung a mask of the burgomaster's face. And then the whole disfigured mess was hanged by order of the court. The pelt of the transmogrified wolf was stuffed and preserved as a proof to the skeptical of the existence of werewolves.

These examples I selected more or less at random from hundreds in a wonderful book published in 1906 by E. P. Evans. These were not isolated cases. This is how animals were once perceived—as potentially culpable beings. Some hint of how widespread such punishment of animals was comes from a passing comment that Gratiano makes while robustly damning Shylock in the *Merchant of Venice*:

> thy currish spirit
> Govern'd a wolf, who, hang'd for human slaughter,
> Even from the gallows did his fell soul fleet,
> And, whilst thou lay'st in thy unhallow'd dam,
> Infused itself in thee;

If there was anything at all uncommon about "a wolf . . . hang'd for human slaughter" Shakespeare would surely have expanded the allusion. But no, for Elizabethan audiences this was clearly understood as a perfectly natural thing to do.

These things happened in America too. The twenty people who were hanged or crushed to death in Salem, Massachusetts, are quite well known. Few people know that two dogs were also hanged for witchcraft during that crazy four-month period in 1692. There is an ongoing battle to clear the names of some of those hanged as witches who have still not been exonerated. Nobody seems interested in exonerating the dogs.

In the debate about animal rights nobody wants to raise the specter of animal responsibilities. As my right to swing my arm stops at my neighbor's nose, so rights in general imply responsibilities. And responsibilities demand comprehension: a defendant must be able to understand what they have been charged with. To express one's rights and accept one's responsibilities one needs a comprehension of others' motivations—a theory of mind. Do any nonhuman species possess the mental wherewithal to meaningfully have rights and responsibilities? The arguments for ape legal rights generally discuss only the possibility of the ape as the aggrieved party, but there is absolutely no grounds for this assumption. Apes often harm each other (Jane Goodall's justly celebrated account of the lives of the chimpanzees at Gombe, *The Shadow of Man*, is replete with tales of chimpanzee battles), and at least two celebrated apes have killed or injured human beings. Savage-Rumbaugh's star bonobo, Kanzi, has injured several people. And one of Jane Goodall's chimps, Frodo (yes, she named him after a hobbit), killed a local toddler in Gombe.

So what is the evidence that apes, or any other nonhuman species, share enough of our psychology to enjoy rights and be burdened with responsibilities? At the outset I said I thought that, in comparing the psychology of other species to our own, I could see a similarity sandwich. Now that we have surveyed a range of animals and a range of psychological skills, let's cut into the similarity sandwich in more detail.

SIMILARITY SANDWICH: THE BOTTOM LAYER

On the bottom, in a layer where all species are distinguishable—human from honeybee, aardvark from giraffe—are our perceptual and sensory abilities. Here we distinguish the dog's world of smells from the bat's echo world; the pigeon's six channels of color vision from the cat's blurry, monochrome landscape. Here, diversity is the norm; nobody is surprised or shocked by it. Different species live in varied niches, leading multifarious lives, so we do not expect them all to see, smell, and hear in similar ways. And this expectation has been abundantly confirmed.

We smell—and we can smell. Though not our favorite sense for everyday navigation, powerful smells can be powerful signals. Death and disease; sex and sweat; children, flowers, and sugar all have their characteristic odors. Some smells send a pretty imperative signal, but no smell influences us as powerfully as the pheromone odors that drive honeybees to give their lives by stinging where others before them have stung, or compel worker bees to forgo egg laying and instead tend the young of their queen. Bees even smell in stereo, with a different sensor on the end of each antenna, so that they can track down the source of an odor. Both bees and wasps are able to sniff out the subtle smell of TNT that seeps from landmines—if only we too could do that, think how many lives and legs could be saved.

Pigeons take their sense of smell further: they use the characteristic odor of home to achieve the (to my mind) magical feat of flying from a site they have never visited before back to a home loft that lies beyond the horizon. Another skill that could save human lives, if only we shared it.

But pigeons don't home only by smell; they use their eyes too. What we see we call visible light. Pigeons' eyes also see what we call ultraviolet, and many insects (including honeybees) see it too. Ultraviolet light is so useful—for spotting texture and pattern in flowers, for finding the sun behind clouds—that it is something of a puzzle that we can't see it too. But in evolutionary terms we are lucky to have any color vision at all. Most mammals don't, because early in the evolutionary history of mammals, they were all nocturnal and had no use for such a luxury. Only our family, the primates, developed color vision, and even ours is not as fancy as that of pigeons. The vision that pigeons (and probably many other birds) have enables them to find patterns where we see none. They also see polarization in patches of blue sky that produce no sensation in us, and from these patterns of polarization they deduce the position of the sun and hence compass directions.

Our hearing can also be bettered. As we have seen (or heard), bats and dolphins use sound not just to locate noisy objects in their environments but also to pinpoint the silent. In both bats and dolphins echolocation is an adaptation to an environment where the distance senses more commonly used by mammals, like vision and smell, are inadequate to the task of finding prey. Echolocating animals are not a homogeneous group. No dolphin could make any use of the sounds made by a bat, or vice versa. For that matter, different species of bat cannot use each other's echo sounds. Bats can localize tiny moths in three dimensions and also estimate the time till contact. The echolocation system of most bats offers finer resolution in time and space than that of dolphins, but dolphins, because of the greater impedance that water offers to sound, are able to use their sonar system to probe the internal structure of objects.

SIMILARITY SANDWICH: THE FILLING

Above all this wonderful diversity, the whirring, humming, perceiving, reacting multiplicity of animal nature, is another more peaceful layer. The layer of the similarity sandwich where we find things that many species, perhaps all, have in common. Not, I suggested, a firm, slice-of-ham type of layer but a squidgy, ambiguous zone: more like peanut butter or cream cheese that oozes into the bread. Here we see animals reacting to signals that predict pain and pleasure: the click of a broken branch that indicates the approach of a dangerous predator; the characteristic raised-tail posture, the glimpse of reddish breast feathers, or whatever it is that indicates that now is a good moment to push for your conjugal rights.

We saw in chapter 3 that wasps are not only capable of a kind of simple learning about cues and consequences (associative learning) but that they also come equipped with numerous patterns of behavior driven by instinct. The term instinct has drifted in and out of scientific respectability, but it seems like a good idea to have a word for behaviors that most members of a species carry out in a similar way. The label "instinct" fell out of fashion for a while because it was taken to mean that the behavior might be entirely genetic, without any environmental component. Which, of course, is silly. There is no behavior that will show itself without the appropriate environment to support it. How can a wasp provision its egg with stunned prey if there is no suitable prey, no substrate for the burial of the egg and prey, and so on? Each species has its own instincts, but every species has some instincts. Consequently, instincts are part of the filling of the similarity sandwich: something all creatures share.

Other candidates for the filling are hard to be sure of. The study of animal behavior is such a patchy affair that today's declarations about skills that all animals have could be proved wrong tomorrow. Still, we could venture some possible similarities worth exploring.

All animals tested show a sense of time, both the time of day and also the length of brief and arbitrary time intervals on the scale of seconds to minutes. Some sense of number also appears widespread, at least for small quantities up to about seven. We saw, for example, in chapter 3 how Otto Koehler's raven Jakob could select a pot with five spots on its lid out of five pots with different numbers of spots on their lids, even though the sizes of the spots varied fifty-fold.

Navigational skills have also been demonstrated in a wide range of species. We talked about how pigeons find home when released from unfamiliar sites hundreds of miles away by exploiting their sense of smell, the polarization of sunlight, and a magnetic sense. We also considered how honeybees track back to the hive from their flower-foraging trips through dead reckoning based on the flow of objects past them as they fly and the memory of landmarks along the route.

Memory in general seems to be a skill that all species possess, at least to some degree. Just how similar memory is in different species it is too early to say, but all animals appear to be able to use past experience to guide present behavior in some way. In the laboratory pigeons can remember which out of hundreds of arbitrary visual patterns will be followed by food, and their memories show little sign of degradation months after the initial experiment. Pigeons also remember what their neighborhood looks like, so that they can find their own loft as they return from homing flights. Honeybees remember where the good flowers were and how to get to them. Rats can remember which parts of a maze contain food. Chimpanzees in the wild can remember where they left the good heavy stones that make excellent anvils for bashing nuts open. Chimpanzees in the laboratory can remember the correct order to press a series of numerals on a computer screen in order to obtain a food treat. Vampire bats can remember who has given them a blood donation in the past and use that information in deciding whether to respond to a petitioner who is begging for a little blood. Memory is widespread.

As we saw in chapter 3, the problem-solving abilities that we call reasoning are more common than one might have expected. Pigeons, rats, squirrel monkeys, and chimps all form inferences of certain types (though not of others). According to certain senses of the word "reasoning," we could count the wasp's ability to learn the smells that reveal where caterpillars are munching on plants as reasoning. Certainly the ability of chimps, capuchin monkeys, and many other species to use tools counts as reasoning. It's really too early to say whether reasoning is common to all animals, but hints of it are certainly present in a wide range of species.

SIMILARITY SANDWICH: WHAT'S ON TOP?

But I am convinced that something belongs on top of the sandwich. There is a layer of characteristics that humans and other species do not have in common. Some of the things we do are unique to us. Language is uniquely human, as is that sense of ourselves as independent beings that we call self-awareness, and there may be other capacities as well. I don't think the number of psychological qualities we ascribe to the top layer is critical, but it is important to recognize that this layer exists.

In Prokofiev's *Peter and the Wolf*, a bird asks a duck, "What kind of bird are you if you can't fly?" To which the duck replies, "What kind of bird are you if you can't swim?" There really are things that some species can do and others can't. Since we are human beings, we are particularly interested in the things that people can do that other species can't. As we saw in chapter 5, there is one ability that stands out above all others as unique to humans: language. My suspicion is that the other things we think make our psychology unique, such as self-awareness, may be consequences of language.

I know that other species communicate. In chapter 2 we discussed the most complex form of natural communication in the world aside from this language you and I are using now. Honeybees communicate to their hivemates the distance, quality, and sun bearing of a nectar source by dancing. This is an astonishing and wonderful achievement, but not one to be confused with human language. With language my plumber can tell me where the pipes are hidden that bring water to the bathroom, and I can tell you that he can tell me that. These are practical achievements of immense import. Compared to my plumber's skill with English, Kanzi and the other chimps are not just tongue-tied; they are dumb. Neither Kanzi nor any other language-trained ape shows any appreciation of grammar (their vocabularies are also extraordinarily constrained). Without grammar there is no language. There can be some primitive communication, but it is communication restricted to demanding objects. And that's it. Sue Savage-Rumbaugh likes to point to Kanzi's tendency to utter certain tokens in set orders on those very rare occasions that he bothers pressing more than one button on his keyboard. But grammar is not simply a tendency to prefer one word order over another; it is an ability to comprehend that different word orders produce different meanings.

There are those who would argue that our human capacity for self-awareness is shared by certain great apes and dolphins. If we knew for sure what self-awareness was, we might be able to test this hypothesis more readily. The supporters of the idea that chimpanzees and dolphins are self-aware point to the ability to recognize oneself in a mirror—or at least to point to spots on one's own body that are only visible in a mirror—as a sign of self-awareness. I have already said that I find it difficult to swallow the idea that pointing to spots seen in a mirror is an acid test of self-awareness. I can see reasons having nothing to do with self-awareness why a creature might touch itself after seeing a spot on an object in a mirror. And I can see many good reasons why a creature with self-awareness might *not* point to itself when confronted by a mirror-image that bore a spot. In any case, if dolphins have self-awareness (as some who have performed the mirror test on these animals have claimed), and gorillas and chimpanzees have it too, then why don't all the animals that share a common ancestor with dolphins and apes also have it—animals like monkeys, dogs, and camels?

What we really need here is a sober and systematic inquiry into what animals perceive when they look into mirrors. Unfortunately, in the rush to catalog which species "recognize themselves" in mirrors, methodical study of this necessary preliminary question got swept under the carpet.

So to understand the psychology of other species, it is crucial to acknowledge that there are ways that we humans differ from them. This in no wise amounts to a backdoor readmission of the *scala naturae*, the ladder of nature that worked its way from worms at the bottom to angels at the top, passing insects, mammals, and people on the way up. Sure, I've said that this layer of the sandwich is on top—but you can turn the sandwich upside down if you like, it makes no difference to the logic. For that matter, we can stretch the metaphor: let every species have its own sandwich, each with something unique to it on the top layer. It's just a metaphor; we can push it around any way we want to. Toast it if you like.

I do not mean to suggest that the human is especially endowed: each species has its unique adaptations to its world. Ours are not "better" than theirs. Imagine the top layer of a bat's sandwich. Here we do not find language or self-awareness but a finely tuned system for catching insects on the wing in pitch-black darkness. An anthropomorphized bat might note that humans are notoriously ineffective at catching flying insects, even in broad daylight, as evidenced by the mosquito bites that mark our limbs. If we were examining the top layer of a vampire bat's similarity sandwich, we would note an ability to stealthily land on a sleeping mammal, bite gently, and lap up its blood without waking it. The top layer of a honeybee's similarity sandwich would include dancing to hivemates to indicate sources of nectar and a total commitment to the welfare of the hive equal to that of a suicide bomber. The top layer of the dolphin's sandwich would include sonar that penetrates solid objects under water, and the ability to control one's breathing to such an extent that dives of hundreds of feet are routine.

We are not magical beasts but ordinary, mortal beings. Our history has led to the development of certain skills, of which language stands out as especially important in understanding what makes us different. If a bat were writing this book, it could reasonably emphasize echolocation as its unique behavioral adaptation. But we shouldn't shrink from recognizing that a bat wouldn't write this book. It is the possession of language, self-awareness, and an awareness that others have minds too that makes it plausible for us to have rights and responsibilities. Absent these qualities, arguing for animal rights makes as much sense as advocating bicycles for fish.

WHAT WOULD DARWIN DO?

There are those—Sue Savage-Rumbaugh, Roger Fouts, Stephen Wise, Jane Goodall among them—who argue that it is un-Darwinian to point out mental faculties that only humans possess.

They emphasize the great and real psychological similarities that exist between humans and other species, especially our closest cousins, the other great apes. They also point to the genetic similarity between chimpanzees and us and argue that this genetic similarity demands fundamental identity in our mental lives.

Nothing Darwin wrote indicates that he thought that people shared all of their psychology with any other species. Diversity was what Darwin saw when he looked at the species on this planet. In *The Descent of Man* and *The Expression of Emotions in Animals and Man*, as well as in unpublished notes, he looked for evidence of continuity between animals and man. He suggested continuity in the expression of emotions, for example, and predicted that we would notice "some of the same general instincts, and feelings" between the other animals and us. But he never suggested identity.

Charles Darwin, by the way, combined enthusiastic support for hunting and a vocal defense of vivisection with a trenchant critique of cruelty to animals. His son recounts how Darwin senior would berate those who treated their horses harshly, and his sympathy with "the educational miseries of dancing dogs."

I have already (in chapter 7) discussed the arguments for culture and self-awareness in chimpanzees and (in chapter 5) the evidence for language. No amount of genetic similarity can make the findings we covered in these chapters closer to what we observe in human societies. Sure, the use of tools by some nonhuman primates, and the way, in some cases, these can be passed on from individual to individual through a primate society show some glimmers of culture—but it is nothing more than that, just a faint glimmer. It's fascinating and beautiful to find that Japanese macaques wash potatoes (potatoes that for years have been delivered prewashed). It's thrilling to learn that chimpanzees remember where the good heavy stones are that make the best anvils for crushing nuts. But we would be nuts not to see the differences between such feats and human culture. To suggest, as Roger Fouts has done, that "the behavior of wild chimpanzees is not so different from that of nontechnological groups of humans" seems to me nothing short of demeaning to those people who live without the benefit of technology. The difference between the mental lives of humans, even those of us who cannot read and write, and those of other species is undeniable.

But what about our genetic near-identity to chimpanzees? Surely if we share pretty much all our DNA with chimpanzees, then they must have pretty much all the psychological qualities that we have?

If sharing 98 percent of our genetic material implies that chimpanzees share effectively all our behavioral traits and mental skills, then we must be able to do pretty much everything they can do too. So if you believe that chimps and we are genetically as-nearly-identical-as-makes-no-odds, please demonstrate how you can play the piano with your toes. Now swing downtown by jumping across rooftops (you may use the higher branches of trees to cross streets). Only when you have done that can you come back and tell me that chimpanzees share every significant aspect of human psychology. Oh yes, and I assume that, as a good chimpanzee should, you murdered your second wife's children by her first husband when you married her.

There are so many things wrong in arguing from percentages of DNA to psychology that it is hard to know where to start.

One good thing to get straight first is a basic understanding of the code in which our genetic material is written. In the DNA code there are only four "letters": the four amino acid bases which code, through their different combinations, for the proteins that build all of life on earth. The fact that there are only four letters in this code means that no two life forms can share less than 25 percent of their DNA (just as a student cannot score below 25 percent on a four-alternative multiple choice exam). In fact, since all forms of life on earth share a common ancestor (as far as we know, life only originated once on this planet), the minimum amount of DNA that any two randomly chosen life forms on earth might share is probably higher than 25 percent. In a fascinating exploration titled simply *What It Means to Be 98 Percent Chimpanzee*, molecular anthropologist Jonathan Marks suggests that we probably share around 35 percent of our DNA with a daffodil. So "percentage of common DNA" is a misleading scale. It doesn't climb from zero (no similarity in form or functions) to 98 or 99 percent (pretty much identical). On the contrary, Marks points out how this brute-force measurement of percentages of shared DNA fails to capture numerous ways in which chimpanzee and human genetic material vary. Chimpanzees do not even have the same number or structure of chromosomes (the packages into which DNA is bundled) as humans. The high figure we get for the similarity of human and chimpanzee DNA is, ultimately, just a function of the immense period of time that there was life on earth prior to the development of modern humans and chimpanzees. In fact, considering that 99.9 percent of the 3½ billion years that life existed on earth passed prior to the separation of humans and chimpanzees from a common ancestor, it seems that the 98.4 percent of DNA that we have in common is, if anything, a little lower than we might expect.

The apparent precision of the 98.4 percent figure masks our ignorance of how to handle numbers like this. Two people, or two chimps, can be genetically identical and yet still lead different lives (I'm thinking of identical twins). One twin might learn French, the other not. One might die young of a freak illness that sweeps through the forest; the other might go on to have children and grandchildren. The method that is used to arrive at the figure of 98.4 percent identical DNA is a method by which any two human beings (or any two chimps) would appear to be at least 99.9999 percent identical. You and I would appear completely indistinguishable. And yet we know that different people (and different chimps) vary tremendously in their use of language, culture, and all those other things that make each species unique.

The fundamental point is this: we cannot predict from percentages of shared DNA what psychological qualities two individuals will have in common. If we want to know about the psychological differences and similarities between two species, then we must study their respective abilities directly. No other form of study, no matter how technical and apparently precise, can stand proxy for research into animal psychology.

CAN THEY SUFFER?

If we are looking for something about animals that compels us to protect them, then I think the position of Jeremy Bentham, English philosopher of the late eighteenth and early nineteenth centuries, is more cogent than the argument that animals possess language and self-awareness: "the question is not, Can they *reason*? Nor, Can they *talk*? But, Can they *suffer*?"

I have met Jeremy Bentham. In a manner of speaking. Bentham founded my alma mater, University College London, and he decreed in his will that his skeleton, dressed in his clothes and with his preserved head on top, should be kept in an open box at the college and wheeled into committee meetings, where the minutes record "Jeremy Bentham—present but not voting." The official University College web site phlegmatically dismisses any claim that Bentham played a role in the founding of the college, but the presence of an early nineteenth-century English gentleman in a glass-fronted wooden case in the South Cloisters is undeniable.

It is Bentham's philosophy, known as utilitarianism—we should seek the greatest happiness of the greatest number—that inspires Peter Singer, progenitor of the animal liberation movement. Singer, in his *Animal Liberation*, the "bible of the animal liberation movement," takes as his starting point less the greatest amassment of happiness than the greatest avoidance of pain. According to Singer, we should strive to reduce the pain in the world. Animals feel pain; therefore we should avoid inflicting hurt on them, no less than we avoid inflicting pain on fellow human beings.

The first question to address here is how do we know that other species feel pain? Singer says we know this in just the same way as we know that other members of our own species feel pain: we seem them writhe, cry out, and attempt to escape things that we too would find painful.

I think this is a much better argument than the one that says we should give apes the same rights we give humans because they are self-aware and use language. First, an attitude toward animals built around avoiding pain could be of much more help to many more species; and second, it is at least a plausible deduction from the available evidence about other species that they may feel pain.

On balance, however, it is not a philosophical resting place with which I (admittedly a nonphilosopher) can be entirely comfortable.

For one thing I don't see how we can perform the complicated calculus needed to ensure that our actions toward animals do indeed lead to a minimization of pain in the world. Sure, beating your dog pointlessly and vindictively cannot do other than add to the world's total of pain, but then nobody's philosophy tolerates such behavior. The Bible, which gives man dominion over the animals, does not endorse wanton cruelty toward them. But if we are concerned about whether or not to use chimpanzees in research on hepatitis, we would need to run a very complex computation to figure out whether the total pain in the world would be increased or decreased by such research. We would need to estimate the pain to each chimpanzee from the research it participates in, then calculate the pain of each human being who is infected with hepatitis (and of those who see their loved ones suffering with the disease). Presumably these calculations must be scaled by the number of chimpanzees used in the research, the probability of finding a cure, the number of people who catch hepatitis, the probability that a cure—if available—would actually reach them, the possibility of finding a cure through research that didn't involve animals, and so on and so forth. All these quantities would somehow have to be estimated before we could reckon whether using chimps in this research was justified. I don't see how these calculations could be done. Furthermore, I do not think that we can in principle know how much future pain or pleasure our present actions may evoke. Nature is far too fickle a mistress for minimizing pain or maximizing pleasure to be a predictable outcome except in the most trivial cases.

Singer's approach is to point to perverse acts of sadistic cruelty toward animals and then argue that these make it clear to any reasonable person how we could treat animals better. But this is to take the easy way out. We are all against wanton cruelty. Descartes, who was the first person to suggest that animals are mere machines, did not (despite what is often claimed) endorse cruelty toward animals. He had a pet dog, Monsieur Grat, of whom he was very fond, and at certain points in his writings he talks of the "feelings" and "desires" of animals.

At one point only in his book does Singer acknowledge that reducing the total pain in the world might be a tricky thing to do. He considers the (obviously very weak) criticism of his position that since the world contains carnivorous animals, we human beings can justify our ill-treatment of animals as something "natural":

> It must be admitted that the existence of carnivorous animals does pose one problem for the ethics of Animal Liberation, and that is whether we should do anything about it. Assuming that human beings could eliminate carnivorous species from the earth, and that the total amount of suffering among animals in the world would thereby be reduced, should we do it?
>
> The short and simple answer is that once we give up our claim to "dominion" over the other species we should stop interfering with them at all. We should leave them alone as much as we possibly can. Having given up the role of tyrant, we should not try to play God either.

He goes on to consider an example of a beneficial human intervention in the natural world (the release of two whales trapped in Arctic ice) before concluding "except in a few very limited cases, we cannot and should not try to police all of nature."

Here Singer seems to be acknowledging that the application of a calculus of pain is far trickier than he has elsewhere admitted. Surely, if pain minimization were our overriding concern, then extinguishing carnivores from the planet would be an entirely consistent action. I could do penance for the hamburger I had yesterday by shooting a couple of eagles. Removing carnivores stands as good a chance of reducing pain in the world as any other action I might engage in.

Let me ask again, how do we know that animals feel pain? Singer's answer, the standard answer, is because they act as we do when we feel pain. Some of the time at least. But pain and signs of pain are not always clearly connected. Our dog Benji was an amazing stoic. He walked into the side of a parked car once, whacking his head against the bodywork with a loud thump. Didn't seem to bother him at all.

When organs are removed from the brain dead for transplantation, doctors commonly give anesthetics. Why bother with anesthetics if there is no chance that the individual is conscious? Because without them the body reacts violently: "You stick the knife in and the pulse and blood pressure shoot up." reported a surgeon interviewed by the BBC. "Nurses get really, really upset." So this adds a further complication to the calculus of pain that Singer wants us to engage in: outward signs may correlate little with inner agonies.

Even if we could measure pain in others, make reasonable estimates of the future pain our present actions may impose, and perform the calculus of pleasure and pain that utilitarians want us to, it is still not clear that this would tell us what to do and to whom. The followers of Leopold von Sacher-Masoch (after whom masochism is named) would fit into an odd class in this regard. Is their pleasure pain? Or their pain pleasure? Or both? More generally, it does not make sense to say that the painfulness of a pain is all that matters about it.

Suppose we have solved the problem of measuring pain in others, both other people and other species. Let's say we've devised a 10-point scale on which any individual's pain can be measured with confidence. And the buck stops there: we assume there are no consequent pains or pleasures in others that follow from one individual's agonies. Let us say I am confronted with a situation where my infant son might suffer a mild pain (say, 2 on our scale), or some other person (the plumber perhaps) must suffer a stronger pain (say, 4 points). I get to choose. Well, I'm sorry, Mr. Plumber, but I vote that you get a 4-point pain. Now I have to choose whether the plumber suffers a 4-point pain or the dog gets a 6-point pain. Much as I love the dog, I'm going to vote that he gets the pain. I do not just care how much pain is in the world: I care about that pain in relation to myself. I love my son more than I love the plumber. (He's a decent guy—just not my type.) And, though I have stronger feelings for the dog than for the plumber, I just have the sense the dog can take pain better than the plumber. At least the dog won't fret over it afterward. In his world the pain will sooner be forgotten. There is nothing intrinsically unreasonable or evil in admitting that some pains matter to me more than others. Philosophers express this by saying that different pains may have different moral status.

So on balance a utilitarian account of animal suffering can tell us not to inflict pain where there is no possible hope of a benefit—something no thoughtful person advocates in any case—but it cannot take us further than that. It cannot form a useful basis for deciding what to do about animals in agriculture or medical research.

Do you think I'm nit-picking? Do you perhaps think that in any real case you'd be able to tell whether any action toward an animal would lead to a greater or lesser amount of pain in the world?

Stephen Budiansky, in *The Covenant of the Wild*, considers the action of adopting a cat from the neighborhood cat shelter. The ads in our local paper for pet adoption are headed "Death Row"—they show photographs of animals who will be euthanized if not adopted. What could be a more gracious and charitable act than saving a poor cat from lethal injection? As Budiansky puts it, "Here is a clear-cut choice between life and death, between kindness and neglect. No animal rights activist would fail to applaud." Well, there is a hidden calculus of suffering here that makes kitten adoption look like a much less benevolent activity. Peter Churcher and John Lawton studied the impact of cats on the local wildlife in a village in Bedfordshire, England. They asked the villagers to collect all the prey their cats brought home in one year. The seventy-eight cats in this village collected 1,101 animals. One-third of all the house sparrows in the village ended up in their maws. Other researchers have found that cats only bring home about half of their catch. On this basis Churcher and Lawton calculated that the 5 million cats in the United Kingdom kill 70 million creatures per year. In the United States 65 percent of all domestic cats are allowed to roam outdoors—that's around 47 million animals. A roughly equal number of wild cats, descended from pets, also roam the nation. Together these 100 million felines destroy about a billion small mammals and hundreds of millions of birds each year. As Churcher and Lawton put it, "Beware of well-fed felines."

I have said that I do not see anything irrational in giving different weight to different pains. I concern myself with my infant son's pains more than the same, or greater, pain in other people: Is this discrimination? Sure it is. Just as I discriminate in his favor when I spend my money on his clothes and comfort and not on that of the poor of this parish. Furthermore, I doubt if many people would reproach me for this. Certain forms of discrimination are rational, and natural.

Both Bentham and Singer have compared the special consideration we give fellow members of our own species to slavery, racism, and sexism. Singer calls this "speciesism." Frankly, I find this comparison demeaning to enslaved people, nonwhites, and women. Discriminating in law between different classes of people is wrong because there is nothing about different types of persons that can justify it. On the other hand, there are numerous differences between us and other species that can justify different treatment before the law. As John Quincy Adams responded in the House of Representatives to a wit who suggested that presenting to the House a petition from slaves would lead to petitions from animals, "Sir . . . if a horse or a dog had the power of speech and of writing, and he should send [me] a petition, [I] would present it to the House." I would remove the requirement that the horse or dog write down and post its petition: I would be happy to accept its testimony verbally.

ANIMAL RIGHTS AND WRONGS

Animals don't need legal rights to be accepted as valuable and worthy of our protection. They are valuable to us because of who we are, not what they are. Things don't have to be like us to be important to us. They don't have to be conscious to be valuable. They don't have to be able to feel pain to deserve our concern. On occasion we value inanimate objects even more than human life.

In 1993 a bomb at the world-famous Uffizi Gallery in Florence destroyed some irreplaceable works of art and damaged several more. This was part of a campaign by the Sicilian Mafia to destabilize the Italian government through "cultural terrorism." Along the way five people were killed—but who cares? It was the destruction of inanimate art works that grabbed world attention, not the slaughter of innocents (unless you happened to know one of these unfortunates personally).

Before the events of September 11, 2001, catapulted the Taliban into world consciousness, their major claim to infamy was that they destroyed two giant Buddhas carved into the cliffs of Bamiyan in the third century A.D. These were the tallest standing Buddhas in the world. Though the Taliban had been ruling Afghanistan with a rod of iron for years, it was the threat to demolish the Buddhas that brought a delegation from the Organization of the Islamic Conference to Kandahar. U. N. Secretary-General Kofi Annan also urged the Taliban to hold back, and Egypt's President Hosni Mubarak sent out the mufti of the republic, his country's supreme Muslim cleric, to try to get the Taliban to spare the statues. All to no avail. "The destruction work is not as easy as people would think," Taliban Information Minister Qudratullah Jamal boasted. He might be forgiven for being confused as to what might impress the international community. After all, the Taliban never heard a whimper of concern from beyond the borders of Afghanistan when they massacred the local people who lived around the Buddhas. How can we be so callous as to ignore the deaths of innocent people and yet be concerned at the damage to mere stones? Sometimes we just do care about inanimate things.

The presence of animals in our world is a thing of very great value. The opportunity to observe and study animals is a source of wonder and excitement, both scientific and aesthetic. There can be no doubt that we must treat these beings with care and respect, just as a student of history must treat the fragile and irreplaceable documents she is given access to with care and respect.

Animals don't have to be like us to be valuable. We don't have to pretend that some species have consciousness equivalent to ours. They don't, and they don't need it to matter to us and deserve our protection. Nor does the ability to feel pain offer us a straightforward way of deciding which kinds of actions toward animals are appropriate and which are not. If explorers in the Amazon basin discover tomorrow the first animal ever seen that shows no signs of pain, we should cherish that animal. We should not, as would Singer presumably, dismiss this beast from our realm of concern because of its inability to feel pain. On the contrary, its exceptional psychology would make it all the more fascinating and in need of our protection. It's not their like-us-ness that makes animals important: It's their not-like-us-ness that is the better reason to cherish them.

You may disagree. I respect that. Thoughtful people can in good faith reach different conclusions on these problems. Let's not bomb each other's childhood haunts. Go peacefully—and may your dog go with you.

FURTHER READING

Animal Liberation, by Peter Singer (HarperCollins, 2002). The original "bible" of the animal liberation movement, first published in 1975. Most of the argument is in chapters 1 and 6.

The Great Ape Project: Equality beyond Humanity, edited by Paola Cavalieri and Peter Singer (St. Martin's, 1994). This manifesto of the Great Ape Project includes supportive contributions by numerous significant figures such as Richard Dawkins, Jane Goodall, Douglas Adams, Jared Diamond, and Francine Patterson.

Rattling the Cage: Toward Legal Rights for Animals, by Steven M. Wise (Perseus Books, 2001). Wise picks up the baton from The Great Ape Project and argues in more detail that chimpanzees deserve the same rights as humans.

The Covenant of the Wild: Why Animals Chose Domestication, by Stephen Budiansky (Yale University Press, 1999). In this very striking book, Budiansky argues that farm animals chose their captivity. A very thought-provoking read.

Animal Cognition: The Mental Lives of Animals, by Clive D. L. Wynne (Palgrave, 2002). If you want more animal psychology and less philosophy than you got in the present study, then my college textbook may be for you.

What It Means to Be 98 Percent Chimpanzee: Apes, People, and Their Genes, by Jonathan Marks (University of California Press, 2002). This book delivers exactly what its title promises: it explains what we can and can't deduce from the fact that we share 98 percent of our DNA with chimpanzees.

The Criminal Prosecution and Capital Punishment of Animals, by E. P. Evans (Lawbook Exchange [Union, N. J.], 1998). Originally published in 1906, this is a thorough and deeply troubling survey of the practice of trying and punishing animals in medieval Europe.

The Philosopher's Dog, by Raimond Gaita (Text Publishing [Melbourne], 2002). Gaita's careful, sympathetic, and thoughtful book is a great antidote to the shrill pronouncements of so many animal advocates.

References

Chapter 1

Animal activist dies on hunger strike. 2001. *British Broadcasting Corporation Online Network*. November 5. <www.bbc.co.uk>.

Barrett, P. H., P. J. Gautrey, S. Herbert, D. Kohn, and S. Smith eds. 1987. *Charles Darwin's Notebooks 1836–1844*. Cambridge: Cambridge University Press.

Bennett, W. 1997. Animal activist "attacked shops with fire-bombs." *Telegraph*, 894 (November 4).

Bruxelles, S. de. 1997. Record sentence for animal rights bomber; Trial. *Times*, December 6, p. 15.

Budiansky, S. 2000. *The Truth about Dogs*. New York: Viking.

Cavalieri, P., and P. Singer, eds. 1994. *The Great Ape Project: Equality beyond Humanity.* New York: St. Martin's.

Dennett, D. 1995. *Darwin's Dangerous Idea: Evolution and the Meanings of Life*. New York: Simon & Schuster.

Griffin, D. R. 2001. *Animal Minds: Beyond Cognition to Consciousness*. Chicago: University of Chicago Press.

Hauser, M. 2001. *Wild Minds*. New York: Henry Holt.

Haynes, K. F., K. V. Yeargan, and C. Gemeno. 2001. Detection of prey by a spider that aggressively mimics pheromone blends. *Journal of Insect Behavior* 14: 535–44.

Joseph, C. 1998. Testing dilemma for McCartneys. *Times*, October 23, p. 8.

McCartney told "animal testing is vital." 1998. *British Broadcasting Corporation Online Network*. Oct 23. <www.bbc.co.uk>.
Moyal, A. M. 2001. *Platypus: The Extraordinary Story of How a Curious Creature Baffled the World*. Washington: Smithsonian Institution Press.
Nagel, T. 1974. What is it like to be a bat? *Philosophical Review* 83: 435–50.
North American Animal Liberation Front Press Office. 2002. *2001 Year-End Direct Action Report*. Available by writing to P. O. Box 3673, Courtenay, B.C., Canada, or at <www.animalliberation.net/index.shtml>.
Ritvo, H. 1987. *The Animal Estate*. Cambridge: Harvard University Press.
Serpell, J. 1996. *In the Company of Animals: A Study of Human-Animal Relationships*. Rev. ed. Cambridge: Cambridge University Press.
Singer, P. 2001. Heavy petting. *Nerve.com*. March 1.
Vigil marks hunger striker funeral. 2001. *British Broadcasting Corporation Online Network*. November 16. <www.bbc.co.uk>.

Chapter 2

Agosta, W. C. 1992. *Chemical Communication*. New York: W. H. Freeman.
Backhaus, W., A. Werner, and R. Menzel. 1987. Color vision in honeybees: Metric, dimensions, constancy, and ecological aspects. In R. Menzel and A. Mercer, eds., *Neurobiology and Behavior of Honeybees*, 172–90. Berlin: Springer Verlag.
Brückner, D., and W. M. Getz. 1991. Odour perception as related to kin recognition. In L. J. Goodman and R. C. Fisher, eds., *The Behaviour and Physiology of Bees*, 60–68. Wallingford, U.K.: CAB International.
Daly, M., and M. Wilson. 1983. *Sex, Evolution, and Behavior*. 2nd ed. Boston: Willard Grant. 30.
Dawkins, R. 1976. *The Selfish Gene*. Oxford: Oxford University Press.
Dreller, C. 1998. Division of labor between scouts and recruits: Genetic influence and mechanisms. *Behavioral Ecology and Sociobiology* 43: 191–96.
Dyer, F. C. 1991. Bees acquire route-based memories but not cognitive maps in a familiar landscape. *Animal Behavior* 41: 239–46.
Esch, H. E., S. Zhang, M. V. Srinivasan, and J. Tautz. 2001. Honeybee dances communicate distances measured by optic flow. *Nature* 411: 581–83.
Getz, W. M., and K. B. Smith. 1983. Genetic kin recognition: Honey bees discriminate between full and half sisters. *Nature* 302: 147–48.
———. 1986. Honey bee kin recognition: Learning self and nestmate phenotypes. *Animal Behaviour*: 34: 1617–26.
Giurfa, M., S. Zhang, A. Jenett, R. Menzel, and M. V. Srinivasan. 2001. The concepts of "sameness" and "difference" in an insect. *Nature* 410: 930–33.
Gould, J. L. 1986. Pattern learning by honeybees. *Animal Behaviour* 34: 990–97.
Griffin, D. R. 2001. *Animal Minds: Beyond Cognition to Consciousness*. Chicago: University of Chicago Press.

Jung, C. G. 1964. *Man and His Symbols*. New York: Dell.

Kevan, P. G. 1987. Texture sensitivity in the life of honeybees. In R. Menzel and A. Mercer, eds., *Neurobiology and Behavior of Honeybees* 96–111. Berlin: Springer Verlag.

Menzel, R. 1987. Memory traces in honeybees. In R. Menzel and A. Mercer, eds., *Neurobiology and Behavior of Honeybees*, 310–25. Berlin: Springer Verlag.

Menzel, R., K. Geiger, L. Chittka, J. Joerges, J. Kunze, and U. Müller. 1996. The knowledge base of bee navigation. *Journal of Experimental Biology* 199: 141–46.

Michelson, A. 1999. The dance language of honeybees: Recent findings and problems. In M. D. Hauser and M. Konishi, eds., *The Design of Animal Communication* 111–31. Cambridge: MIT Press.

Michelson, A., B. B. Andersen, J. Storm, W. H. Kirchner, and M. Lindauer. 1992. How honeybees perceive communication dances, studied by means of a mechanical model. *Behavioral Ecology and Sociobiology* 30: 143–50.

Milde, J. J. 1987. The ocellar system of the honeybee. In R. Menzel and A. Mercer, eds., *Neurobiology and Behavior of Honeybees*, 191–200. Berlin: Springer Verlag.

Moritz, R.F.A. 1988. Group relatedness and kin discrimination in honey bees. *Animal Behaviour* 36: 1334–40.

———. 1991. Kin recognition in honeybees: Experimental artefact or biological reality. In L. J. Goodman and R. C. Fisher, eds., *The Behaviour and Physiology of Bees*, 48–59. Wallingford, U.K.: CAB International.

Moritz, R.F.A., and E. E. Southwick. 1992. *Bees as Superorganisms: An Evolutionary Reality*. Berlin: Springer Verlag.

Neumann, P., C.W.W. Pirk, H. R. Hepburn, A. J. Solbrig, F.L.W. Ratnieks, P. J. Elzen, and J. R. Baxter. 2001. Social encapsulation of beetle parasites by Cape honeybee colonies (*Apis mellifera capensis Esch.*). *Naturwissenschaften* 88: 400.

Noonan, K. C. 1986. Recognition of queen larvae by worker honey bees (*Apis mellifera*). *Ethology* 73: 295–306.

Oldroyd, B. P., T. E. Rindere, and S. M. Buco. 1986. Honey bees dance with their super-sisters. *Animal Behaviour* 42: 121–29.

Oldroyd, B. P., H. A. Sylvester, S. Wongsiri, and T. E. Rinderer. 1994. Task specialization in a wild bee, *Apis florea* (Hymenopter: Apidae), revealed by RFLP banding. *Behavioral Ecology and Sociobiology* 34: 25–30.

Page, R.E.J., and H.H.J. Laidlaw. 1992. Honey bee genetics and breeding. In J. M. Graham, ed., *The Hive and the Honey Bee*, 235–67. Hamilton, Ill.: Dadant.

Page, R.E.J., G. E. Robinson, and M. K. Fondrk. 1989. Genetic specialists, kin recognition and nepotism in honey-bee colonies. *Nature* 338: 576–79.

Seeley, T. 1995. *The Wisdom of the Hive: The Social Physiology of Honey Bee Colonies*. Cambridge: Harvard University Press.

Seeley, T. D. 1997. Honey bee colonies are group-level adaptive units. *American Naturalist* 150, suppl.: S1–41.
Southwick, E. E. 1992. Physiology and social physiology of the honey bee. In J. M. Graham, ed., *The Hive and the Honey Bee*, 171–96. Hamilton, Ill.: Dadant.
Srinivasan, M. V., S. W. Zhang, M. Lehrer, and T. S. Collett. 1996. Honeybee navigation en route to the goal: Visual flight control and odometry. *Journal of Experimental Biology* 199: 237–44.
Stout, J. C., and D. Goulson. 2001. The use of conspecific and interspecific scent marks by foraging bumblebees and honeybees. *Animal Behaviour* 62: 183–89.
von Frisch, K. 1967. *A Biologist Remembers*. Oxford: Pergamon.
Wenner, A. M., and P. H. Wells. 1990. *Anatomy of a Controversy: The Question of a "Language" among Bees*. New York: Columbia University Press.

Chapter 3

Boysen, S. T. 1992. Counting as the chimpanzee sees it. In W. K. Honig and J. G. Fetterman, eds., *Cognitive Aspects of Stimulus Control*, 367–83. Hillsdale, N.J.: Lawrence Erlbaum.
Boysen, S. T., and G. G. Berntson. 1989. Numerical competence in a chimpanzee (Pan troglodytes). *Journal of Comparative Psychology* 103: 23–31.
Brannon, E. M., and H. S. Terrace. 1998. Ordering of the numerosities 1 to 9 by monkeys. *Science* 282: 746–49.
———. 2000. Representation of the numerosities 1–9 by Rhesus Macaques (*Macaca mulatta*). *Journal of Experimental Psychology: Animal Behavior Processes* 26: 31–49.
Burt, C. 1911. Experimental tests of higher mental processes and their relation to general intelligence. *Journal of Experimental Pedagogy* 1: 93–112.
Cheney, D. L., and R. M. Seyfarth. 1990. *How Monkeys See the World*. Chicago: University of Chicago Press.
Darwin, C. 1989. *The Descent of Man and Selection in Relation to Sex*. London: Pickering & Chatto.
Deichmann, U. 1996. *Biologists under Hitler*. Cambridge: Harvard University Press.
Domjan, M. 2003. *The Principles of Learning and Behavior*. 5th ed. Pacific Grove, Calif.: Brooks/Cole.
Fabre, J. H. 1919. *The Hunting Wasps*. London: Hodder & Stoughton.
Griffin, D. R. 2001. *Animal Minds: Beyond Cognition to Consciousness*. Chicago: University of Chicago Press.
Hassenstein, B. 1974. Otto Koehler—sein Leben und sein Werk. *Zeitschrift für Tierpsychologie* 35: 449–64.
Hood, B. M., M. D. Hauser, L. Anderson, and L. Santos. 1999. Gravity biases in a non-human primate? *Developmental Science* 2: 35–41.

Köhler, W. 1925. *The Mentality of Apes*. Trans. E. Winter. London: Kegan Paul Trench & Trubner.

Lewis, W. J., J. O. Stapel, A. M. Cortesero, and K. Takasu. 1998. Understanding how parasitoids balance food and host needs: Importance to biological control. *Biological control* 11: 175–83.

Lewis, W. J., J. H. Tumlinson, and S. Krasnoff. 1991. Chemically mediated associative learning: An important function in the foraging behavior of *Microplitis croceipes* (Cresson). *Journal of Chemical Ecology* 17: 1309–25.

Ley, R. 1990. *A Whisper of Espionage*. Garden City Park, N.Y.: Avery.

Limongelli, L., S. T. Boysen, and E. Visalberghi. 1995. Comprehension of cause-effect relations in a tool-using task by chimpanzees (*Pan troglodytes*). *Journal of Comparative Psychology* 109: 18–26.

McGonigle, B. O., and M. Chalmers. 1977. Are monkeys logical? *Nature* 267: 694–96.

Piaget, J. 1965. *The Child's Conception of Number*. New York: Norton.

Rilling, M. 1967. Number of responses as a stimulus in fixed interval and fixed ratio schedules. *Journal of Comparative and Physiological Psychology* 63: 60–65.

Sheehan, W., F. L. Waekers, and W. J. Lewis. 1992. Discrimination of previously searched, host-free sites by *Microplitis croceipes* (Hymenoptera: Braconidae). *Journal of Insect Behavior* 6: 323–31.

Sparks, J. 2001. Scientists train "sniffer" wasps as mine hunters. *Telegraph*, December 16.

Terrace, H. S. 1993. The phylogeny and ontogeny of serial memory: List learning by pigeons and monkeys. *Psychological Science* 4: 162–69.

Turling, T.C.J., J. H. Loughrin, P. J. Mccall, U.S.R. Rose, W. J. Lewis, and J. H. Tumlinson. 1995. How caterpillar-damaged plants protect themselves by attracting parasitic wasps. *Proceedings of the National Academy of Sciences* 9: 1169–74.

Turling, T.C.J., J.W.A. Scheepmaker, L.E.M. Vet, J. H. Tumlinson, and W. J. Lewis. 1990. How contact foraging experiences affect preferences for host-related odors in the larval parasitoid *Cotesia marginiventris* (Cresson)(Hymenoptera: Braconidae). *Journal of Chemical Ecology* 16: 1577–89.

Visalberghi, E., and L. Limongelli. 1994. Lack of comprehension of cause-effect relations in tool-using capuchin monkeys (*Cebus apella*). *Journal of Comparative Psychology* 108: 15–22.

von Fersen, L., C.D.L. Wynne, J. D. Delius, and J.E.R. Staddon. 1991. Transitive inference formation in pigeons. *Journal of Experimental Psychology: Animal Behavior Processes* 17: 334–41.

Waeckers, F. L., C. Bonifay, and W. J. Lewis. 2002. Conditioning of appetitive behavior in the Hymenopteran parasitoid *Microplitis croceipes*. *Entomologia Experimentalis et Applicata* 103: 135–38.

Wynne, C.D.L., L. von Fersen, and J.E.R. Staddon. 1992. Pigeons' inferences are transitive and are the outcome of elementary conditioning principles: A response. *Journal of Experimental Psychology: Animal Behavior Processes* 18: 313–15.

Chapter 4

Allen, G. M. 1940. *Bats*. Cambridge: Harvard University Press.

Carter, M. L., ed. 1988. *Dracula: The Vampire and the Critics*. Ann Arbor, Mich.: UNI Research.

Dekkers, M. 1994. *Dearest Pet: On Bestiality*. New York: Norton.

Findley, J. S. 1993. *Bats: A Community Perspective*. New York: Cambridge University Press.

Fullard, J. H., J. A. Simmons, and P. A. Saillant. 1994. Jamming bat echolocation: The dogbane tiger moth Cycnia tenera times its clicks to the terminal attach calls of the big brown bat *Eptesicus fuscus*. *Journal of Experimental Biology* 194: 285–98.

Grossetete, A., and C. F. Moss. 1998. Target flutter rate discrimination by bats using frequency-modulated sonar sounds: Behavior and signal processing models. *Journal of the Acoustical Society of America* 103: 2167–76.

Hughes, H. C. 1999. *Sensory Exotica: A World beyond Human Experience*. Cambridge: MIT Press.

Lancaster, W. C., A. W. Keating, and O. W. Henwon Jr. 1992. Ultrasonic vocalizations of flying bats monitored by radiotelemetry. *Journal of Experimental Biology* 173: 43–58.

Nagel, T. 1974. What is it like to be a bat? *Philosophical Review* 83: 435–50.

Neuweiler, G. 2000. *The Biology of Bats*. New York: Oxford University Press.

Robertson, J. 1990. *The Complete Bat*. London: Chatto & Windus.

Roeder, K. D., and A. E. Treat. 1961. The detection and evasion of bats by moths. *American Scientist* 49: 135–48.

Simmons, J. A., M. J. Ferragamo, and C. F. Moss. 1998. Echo-delay resolution in sonar images of the big brown bat, *Eptesicus fuscus*. *Proceedings of the National Academy of Sciences* 95: 12,647–652.

Varma, D. P. 1975. The genesis of Dracula: A re-visit. In P. Underwood, ed., *The Vampire's Bedside Companion*. London: Leslie Frewin.

Waterton, C. 1825. *Wanderings in South America, the North-west of the United States, and the Antilles, in the Years 1812, 1816, 1820, and 1824*. London: J. Mawman.

Wilkinson, G. S. 1984. Reciprocal food sharing in the vampire bat. *Nature* 308: 181–83.

———. 1985a. The social organization of the common vampire bat I: Pattern and cause of association. *Behavioral Ecology and Sociobiology* 17: 111–21.

———. 1985b. The social organization of the common vampire bat II: Mating system, genetic structure, and relatedness. *Behavioral Ecology and Sociobiology* 17: 123–34.

———. 1986. Social grooming in the common vampire bat, *Desmodus rotundus*. *Animal Behaviour* 34: 1880–89.

Chapter 5

Boakes, R. 1984. *From Darwin to Behaviourism: Psychology and the Minds of Animals*. Cambridge: Cambridge University Press.
Bugnyar, T., M. Kijne, and K. Kotrschal. 2001. Food calling in ravens: Are yells referential signals? *Animal Behaviour* 61: 949–58.
Cerutti, D. 2001. E-mail communication with author, August.
Cheney, D. L., and R. M. Seyfarth. 1990. *How Monkeys See the World*. Chicago: University of Chicago Press.
———. 1997. Reconciliatory grunts by dominant female baboons influence victims' behaviour. *Animal Behaviour* 54: 409–18.
———. 1999. Mechanisms underlying the vocalizations of nonhuman primates. In M. D. Hauser and M. Konishi, eds., *The Design of Animal Communication*, 629–44. Cambridge: MIT Press.
Chomsky, N. 1959. A review of Skinner's "Verbal Behavior." *Language* 35: 26–58.
Darwin, C. 1987. *Charles Darwin's Notebooks 1836–1844*. Ed. P. H. Barnett, P. J. Gautrey, S. Herbert, D. Kohn, and S. Smith. Ithaca, N.Y.: British Museum (Natural History) and Cornell University Press.
De Waal, F. 2001. *The Ape and the Sushi Master: Cultural Reflections of a Primatologist*. New York: Basic Books.
Descartes, R. 1976. Animals are machines. In T. Regan and P. Singer, eds., *Animal Rights and Human Obligations*, 60–66. Englewood Cliffs, N.J.: Prentice Hall.
Dunbar, R. 1996. *Grooming, Gossip and the Evolution of Language*. London: Faber & Faber.
Durkin, K. 1995. *Developmental Social Psychology*. Oxford: Blackwell.
Evans, C. S., and L. Evans. 1999. Chicken food calls are functionally referential. *Animal Behaviour* 58: 307–19.
Evans, C. S., and P. Marler. 1994. Food calling and audience effects in male chickens, *Gallus gallus*: Their relationships to food availability, courtship and social facilitation. *Animal Behaviour* 47: 1159–70.
Evans, C. S., L. Evans, and P. Marler. 1993. On the meaning of alarm calls: Functional reference in an avian vocal system. *Animal Behaviour* 46: 23–38.
Gardner, R. A., and B. T. Gardner. 1969. Teaching sign language to a chimpanzee. *Science* 165: 664–72.
———. 1975. Early signs of language in child and chimpanzee. *Science* 187: 752–53.
Gowers, E. 1948. *Plain Words: A Guide to the Use of English*. London: H. M. Stationery Office.

Hauser, M. D., N. Chomsky, and W. T. Fitch. 2002. The faculty of language: What is it, who has it, and how did it evolve? *Science* 298: 1569–79.

Hayes, C. 1951. *The Ape in Our House*. New York: Harper.

Jones, J., writer and director. 1994. Can chimps talk? *Nova*, show no. 2105, February 15. Transcript available at: <www.geocities.com/RainForest/Vines/4451/nova_CanChimpsTalk1.html>.

Kako, E. 1999. Elements of syntax in the systems of three language-trained animals. *Animal Learning and Behavior* 27: 1–14.

Kellogg, W. N., and L. A. Kellogg. 1933. *The Ape and the Child*. New York: Hafner.

Macintyre, B. 1999. Big chimp, small talk. *The Australian*, August 6, p. 13.

Mitani, J. 1994. Ethological studies of chimpanzee vocal behavior. In R. W. Wrangham, W. C. McGrew, F.B.M. de Waal, and P. G. Heltne, eds., *Chimpanzee Cultures*, 195–210. Cambridge: Harvard University Press.

Mitani, J., and T. Nishida. 1993. Contexts and social correlates of long-distance calling by male chimpanzees. *Animal Behaviour* 45: 735–46.

Palombit, R. A., D. L. Cheney, and R. M. Seyfarth. 1999. Male grunts as mediators of social interaction with females in wild Chacma baboons (*Papio cynocephalus ursinus*). *Behaviour* 136: 221–42.

Pinker, S. 1994. *The Language Instinct: How the Mind Creates Language*. New York: William Morrow.

Rendall, D., D. L. Cheney, and R. M. Seyfarth. 2000. Proximate factors mediating "contact" calls in adult female baboons (*Papio cynocephalus ursinus*) and their infants. *Journal of Comparative Psychology* 114: 36–46.

Savage-Rumbaugh, E. S. 1988. A new look at ape language: Comprehension of vocal speech and syntax. In D. W. Leger, ed., *Comparative Perspectives in Modern Psychology* (Nebraska Symposium on Motivation, 1987), 201–55. Lincoln: University of Nebraska Press.

Savage-Rumbaugh, E. S., and K. E. Brakke. 1996. Animal language: Methodological and interpretive issues. In M. Bekoff and D. Jamieson, eds., *Readings in Animal Cognition*, 269–88. Cambridge: MIT Press.

Savage-Rumbaugh, E. S., and R. Lewin. 1994. *Kanzi: The Ape at the Brink of the Human Mind*. Hoboken, N.J.: John Wiley.

Savage-Rumbaugh, E. S., K. McDonald, R. A. Sevcik, W. D. Hopkins, and E. Rubert. 1986. Spontaeous symbol acquisition and communicative use by pygmy chimpanzees (*Pan paniscus*). *Journal of Experimental Psychology: General.* 115: 211–35.

Savage-Rumbaugh, E. S., J. Murphy, R. A. Sevcik, K. E. Brakke, S. L. Williams, and D. M. Rumbaugh. 1993. Language comprehension in ape and child. *Monographs of the Society for Research in Child Development*, no. 58: Chicago: University of Chicago Press.

Seyfarth, R. M., and D. L. Cheney. 1999. Production, usage, and response in nonhuman primate vocal development. In M. D. Hauser and M. Konishi, eds., *The Design of Animal Communication*, 391–417. Cambridge: MIT Press.

Seyfarth, R. M., D. L. Cheney, and P. Marler. 1980. Vervet monkey alarm calls: Semantic communication in a free-ranging primate. *Animal Behaviour* 28: 1070–94.

Silk, J. B., R. M. Seyfarth, and D. L. Cheney. 1999. The structure of social relationships among female savanna baboons in Moremi reserve, Botswana. *Behaviour* 136: 679–703.

Skinner, B. F. 1957. *Verbal behavior*. New York: Appleton-Century-Crofts.

Struhsaker, T. T. 1967. Auditory communication among vervet monkeys (*Cercopithecus aethiops*). In S. A. Altmann, ed., *Social Communication among Primates*. Chicago: University of Chicago Press.

Wallman, J. 1992. *Aping Language*. Cambridge: Cambridge University Press.

Chapter 6

Able, K. P. 1996. The debate over olfactory navigation by homing pigeons. *Journal of Experimental Biology* 199: 121–24.

British Broadcasting Corporation Online Network. 2000. Carla fights Mayor Ken's pigeon plan. October 7. CNN, <www.bbc.co.uk>.

Cher Ami—World War I carrier pigeon. *Encyclopedia Smithsonian*. <www.si.edu/resource/faq/nmah/cherami.htm>.

Cothren, Marion B. 1934. *Cher Ami: The Story of a Carrier Pigeon*. Boston: Little, Brown.

Darwin, Charles, 1859. *The Origin of Species by Means of Natural Selection, or the Preservation of Favoured Races in the Struggle for Life*. London: John Murray.

Delius, J. D., R. J. Perchard, and J. Emmerton. 1976. Polarized light discrimination by pigeons and an electroretinographic correlate. *Journal of Comparative and Physiological Psychology* 90: 560–71.

Dickens, C. 1850. Winged telegraphs. *London Household Word*, February, 454–56.

Eaton, J. M. 1858. *A Treatise on the Art of Breeding and Managing Tame, Domesticated, Foreign and Fancy Pigeons*. London: publ. "for the author."

Farrington, Harry Webb. 1920. *Cher Ami*. New York: Rough & Brown. Available online at <www.longwood.k12.ny.us/history/upton/cher.htm>.

Frontinus, S. J. 1685. *The Stratagems of War*. London: S. Heyrick, J. Place & R. Sare.

Fuller, T. 1651. *The Historie of the Holy Warre*. Cambridge: Philemon Stephens.

Kreithen, M. L., and W. T. Keeton. 1974. Detection of changes in atmospheric pressure by the homing pigeon, *Columba livia*. *Journal of Comparative Physiology* 89: 73–82.

———. 1974. Detection of polarized light by the homing pigeon, *Columba livia*. *Journal of Comparative Physiology* 89: 83–92.

Levi, W. M. 1981. *The Pigeon* 2d ed. Sumter, S.C.: Wendell Levi.

Perusse, D., and L. Lefebvre. 1985. Grouped sequential exploitation of food patches in a flock feeder, the feral pigeon. *Behavioural Processes* 11: 39–52.

Pliny the Elder. 1957. *Pliny's Natural History, compiled from Historia Naturalis.* Ed. Lloyd Haberly. New York: Frederick Ungar.

Rubin, N. A. 1999. The dove. <198.62.75.1/www1/ofm/art/ART9905.html>.

Sheldrake, R. 1994. *Seven Experiments That Could Change the World.* London: Fourth Estate. Further details available at <www.sheldrake.org/experiments/pigeons>.

Skinner, B. F. 1979. *The Shaping of a Behaviorist.* New York: Knopf.

Trafalgar Square pigeons need you!—update. <www.petaeurope.org/alert/tralf.html>.

U.S. Army 77th Regional Support Command [home page]. <www.usarc.army.mil/77thrsc>.

Walcott, C. 1991. Magnetic maps in pigeons. In P. Berthold, ed., *Orientation in Birds*, 38–51. Basel: Birkhauser Verlag.

———. 1996. Pigeon homing: Observations, experiments and confusions. *Journal of Experimental Biology* 199: 21–27.

Wallraff, H. G. 1980. Olfaction and homing in pigeons: Nerve-section experiments, critique, hypotheses. *Journal of Comparative Physiology: A-Sensory Neural & Behavioral Physiology* 139: 209–24.

———. 1990. Navigation by homing pigeons. *Ethology, Ecology and Evolution* 2: 81–115.

———. 1996. Seven theses on pigeon homing deduced from empirical findings. *Journal of Experimental Biology* 199: 105–11.

———. 2000. Simulated navigation based on observed gradients of atmospheric trace gases (models on pigeon homing, part 3). *Journal of Theoretical Biology* 205: 133–45.

Wiltschko, R. 1996. The function of olfactory input in pigeon orientation: Does it provide navigational information or play another role? *Journal of Experimental Biology* 199: 113–19.

Wiltschko, W., and R. Wiltschko. 1991. Magnetic orientation and celestial cues in migratory orientation. In P. Berthold, ed., *Orientation in Birds*, 16–37. Basel: Birkhauser Verlag.

———. 1996. Magnetic orientation in birds. *Journal of Experimental Biology* 199: 29–38.

Chapter 7

Akins, C. K., and T. R. Zentall. 1996. Imitative learning in male Japanese quail (*Coturnix japonica*) using the two-action method. *Journal of Comparative Psychology* 3: 316–20.

———. 1999. Imitation in Japanese quail: The role of reinforcement of demonstrator responding. *Psychonomic Bulletin & Review* 5: 694–97.

Bekoff, M. 2001. Observations of scent-marking and discriminating self from others by a domestic dog (*Canis familiaris*): Tales of displaced yellow snow. *Behavioural Processes* 55: 75–79.

Brown, C., and K. N. Laland. 2002. Social learning of a novel avoidance task in the guppy: Conformity and social release. *Animal Behaviour* 64: 41–47.

Byrne, R. W., and A. E. Russon. 1998. Learning by imitation: A hierarchical approach. *Behavioral and Brain Sciences* 21: 667–84.

Byrne, R. W., and A. Whiten. 1991. Computation and mindreading in primate tactical deception. In A. Whiten ed., *Natural Theories of Mind: Evolution, Development and Simulation of Everyday Mindreading*, 127–41. Oxford: Blackwell.

Curio, E., E. Ernst, and W. Vieth. 1978. Cultural transmission of enemy recognition: One function of mobbing. *Science* 202: 899–901.

Custance, D. M., A. Whiten, and K. A. Bard. 1995. Can young chimpanzees (*Pan troglodytes*) imitate arbitrary actions? Hayes & Hayes (1952) revisited. *Behaviour* 132: 837–59.

Darwin, C. 1989 (1871). *The Descent of Man and Selection in Relation to Sex*. London: Pickering & Chatto.

De Waal, F. 2001. *The Ape and the Sushi Master: Cultural Reflections of a Primatologist*. New York: Basic Books.

Dunbar, R. 1996. *Grooming, Gossip and the Evolution of Language*. London: Faber & Faber.

Galef, B. G. 1990. The question of animal culture. *Human Nature* 3: 157–78.

Gallup, G.G.J. 1995. Mirrors, minds, and cetaceans. *Consciousness & Cognition* 4: 226–28.

———. 1997. On the rise and fall of self-conception in primates. In J. G. Snodgrass and R. L. Thompson, eds., *The Self across Psychology: Self-Recognition, Self-Awareness, and the Self Concept*, 73–82. New York: New York Academy of Sciences.

Goodall, J. 1988. *In the Shadow of Man*. Rev. ed. London: George Weidenfeld & Nicolson.

Guinet, C., and J. Bouvier. 1995. Development of intentional stranding hunting techniques in killer whale (*Orcinus orca*) calves at Crozet archipelago. *Canadian Journal of Zoology* 73: 27–33.

Hare, B., J. Call, and M. Tomasello. 2001. Do chimpanzees know what conspecifics know? *Animal Behaviour* 61: 139–51.

Hare, B., J. Call, B. Agnetta, and M. Tomasello. 2000. Chimpanzees know what conspecifics do and do not see. *Animal Behaviour* 59: 771–85.

Hayes, C. 1951. *The Ape in Our House*. New York: Harper.

Inoue-Nakamura, N., and T. Matsuzawa. 1997. Development of stone tool use by wild chimpanzees (*Pan troglodytes*). *Journal of Comparative Psychology* 111: 159–73.

Köhler, W. 1925. *The Mentality of Apes*. Trans. E. Winter. London: Kegan Paul Trench & Trubner.

Kuo, Z. Y. 1930. Genesis of the cat's responses toward the rat. *Journal of Comparative Psychology* 11: 1–36.

Lefebvre, L. 1995. Culturally-transmitted feeding behaviour in primates: Evidence for accelerating learning rates. *Primates* 36: 227–39.

Matsuzawa, T. 1994. Field experiments on use of stone tools by chimpanzees in the wild. In R. W. Wrangham, W. C. McGrew, P. G. Heltne, and F.B.M. de Waal, eds. *Chimpanzee Cultures*, 351–70. Cambridge: Harvard University Press.

Myowa-Yamakoshi, M., and T. Matsuzawa. 1999. Factors influencing imitation of manipulatory actions in chimpanzees (*Pan troglodytes*). *Journal of Comparative Psychology* 113: 128–36.

———. 2000. Imitation of intentional manipulatory actions in chimpanzees (*Pan troglodytes*). *Journal of Comparative Psychology* 114: 381–91.

Povinelli, D. J., A. B. Rulf, K. R. Landau, and D. T. Bierschwale. 1993. Self-recognition in chimpanzees (*Pan troglodytes*): Distribution, ontogeny, and patterns of emergence. *Journal of Comparative Psychology* 107: 347–72.

Povinelli, D. J., and T. J. Eddy. 1996. What young chimpanzees know about seeing. Monographs of the Society for Research in Child Development, no. 61. Chicago: University of Chicago Press.

Premack, D., and A. J. Premack. 1994. Why animals have neither culture nor history. In T. Ingold et al., eds., *Companion Encyclopedia of Anthropology*, 350–65. London: Routledge.

Premack, D., and G. Woodruff. 1978. Chimpanzee problem-solving: A test for comprehension. *Science* 202: 532–35.

Reaux, J. E., L. A. Theall, and D. J. Povinelli. 1999. A longitudinal investigation of chimpanzees' understanding of visual perception. *Child Development* 70: 275–90.

Reiss, D., and L. Marino. 2001. Mirror self-recognition in the bottlenose dolphin: A case of cognitive convergence. *Proceedings of the National Academy of Sciences* 98: 5937–42.

Ross, J. 1819. *A voyage of discovery, made under the orders of the Admiralty, in His Majesty's ships Isabella and Alexander, for the purpose of exploring Baffin's Bay, and inquiring into the probability of a north-west passage*. London: John Murray.

Russon, A. E., and B. M. Galdikas. 1993. Imitation in free-ranging rehabilitant orangutans (*Pongo pygmaeus*). *Journal of Comparative Psychology* 107: 147–61.

Shakespeare, W. 1997. *Shakespeare's Sonnets*. Nashville, Tenn.: Thomas Nelson.

Thorndike, E. L. 1911. *Animal Intelligence: Experimental Studies*. New York: Macmillan. Available at <psychclassics.yorku.ca/Thorndike/Animal/chap2.html>.

Turner, C. D., and P. Bateson, eds. 2000. *The Domestic Cat: The Biology of its Behaviour*. 2d ed. Cambridge: Cambridge University Press.

Watson, L. 1979. *Lifetide*. New York: Simon & Schuster.
Whiten, A., and D. Custance. 1996. Studies of imitation in chimpanzees and children. In C. M. Heyes and B.G.J. Galef, eds., *Social Learning in Animals: The Roots of Culture*, 291–318. New York: Academic.
Whiten, A., J. Goodall, W. C. McGrew, T. Nishida, V. Reynolds, Y. Sugiyama, C.E.G. Tutin, R. W. Wrangham, and C. Boesch. 1999. Cultures in chimpanzees. *Nature* 399: 682–85.
Woodruff, G., and D. Premack. 1979. Intentional communication in the chimpanzee: The development of deception. *Cognition* 7: 333–62.

Chapter 8

Alpers, A. 1960. *Dolphins*. London: John Murray.
Au, W.W.L. 1993. *The Sonar of Dolphins*. New York: Springer Verlag.
———. 1997. Echolocation in dolphins with a dolphin-bat comparison. *Bioacoustics* 8: 137–62.
Brill, R. L., M. L. Sevenich, T. J. Sullivan, J. D. Sustman, and R. E. Witt. 1988. Behavioral evidence for hearing through the lower jaw by an echolocating dolphin (*Tursiops Truncatus*). *Marine Mammal Science* 4: 223–30.
Caldwell, D. K., and M. C. Caldwell. 1972. *The World of the Bottlenosed Dolphin*. Philadelphia: Lippincott.
Connor, R. C., M. R. Heithaus, and L. M. Barre. 1999. Superalliance of bottlenose dolphins. *Nature* 397: 571–72.
Connor, R. C., R. S. Wells, J. Mann, and A. J. Read. 2000. The bottlenose dolphin. In J. Mann, R. C. Connor, P. L. Tyack, and H. Whitehead, eds., *Cetacean Societies: Field Studies of Dolphins and Whales*, 91–126. Chicago: University of Chicago Press.
Deacon, T. W. 1990. Rethinking mammalian brain evolution. *American Zoologist* 30: 629–705.
Devine, E., and M. Clark. 1967. *The Dolphin Smile: Twenty-nine Centuries of Dolphin Lore*. New York: Macmillan.
Dolphins aid Iraq mine-clearance. 2003. *Cable News Network Online*. March 26. <cnn.com>.
Dolphins help spot mines in Iraq war. 2003. *Salon Online Magazine*. March 26. <salon.com>.
Goffman, O. and K. Lavalli. 1996. Shark attack in the Gulf of Aqaba at Marsa Bareka, Sharm el Sheik and rescue of the swimmer by a pod of bottlenose dolphins, *Tursiops truncatus*, and the crew of the Jadran. *Israel Marine Mammal Research and Assistance Center*. <maritime.haifa.acil/cms/immrac/research/sharks.htm>.
Hooper, J. 1983. John Lilly: Altered States. *Omni Magazine*. January. Available online at www.omnimag.com/archives/interviews/lilly.html.

Hughes, H. C. 1999. *Sensory Exotica: A World beyond Human Experience.* Cambridge: MIT Press.

Iran buys kamikaze dolphins. 2000. *British Broadcasting Corporation Online Network.* March 8. <www.bbc.co.uk>.

Janik, V. M., and P.J.B. Slater. 1998. Context-specific use suggests that bottlenose dolphin signature whistles are cohesion calls. *Animal Behaviour* 56: 829–38.

Jones, S. 2000. *Darwin's Ghost: "The Origin of Species" Updated.* New York: Random House.

Kellogg, W. N. 1961. *Porpoises and Sonar.* Chicago: University of Chicago Press.

Lilly, J. C. 1961. *Man and Dolphin.* Garden City, N.Y.: Doubleday.

———. 1978. *Communication between Man and Dolphin: The Possibilities of Talking with Other Species.* New York: Crown.

Mann, J., R. C. Connor, P. Tyack, and H. Whitehead, eds. 2000. *Cetacean Societies: Field Studies of Dolphins and Whales.* Chicago: University of Chicago Press.

McBride, A. F. 1940. Meet Mr. Porpoise. *Natural History* 45: 16–22.

Pryor, K., and K. S. Norris. 1991. *Dolphin Societies.* Berkeley: University of California Press.

Pryor, K., J. Lindbergh, S. Londbergh, and R. Milano. 1990. A dolphin-human fishing cooperative in Brazil. *Marine Mammal Science* 6: 77–82.

Reynolds, J.E. III,. R. S. Wells. S. D. Eide, eds. 2000. *The Bottlenose Dolphin: Biology and Conservation.* Gainesville: University Press of Florida.

Reynolds, J. E., and S. A. Rommel. 1999. *Biology of Marine Mammals.* Washington, D.C.: Smithsonian Institution Press.

Ridgway, S. H. 1986. Physiological observations on dolphin brains. In R. J. Schusterman, J. A. Thomas, and F. G. Wood, eds., *Dolphin Cognition and Behavior: A Comparative Approach,* 31–59. Hillsdale, N.J.: Lawrence Erlbaum.

Ridgway, S. H., B. L. Scronce, and J. Kanwishe. 1969. Respiration and deep diving in bottlenose porpoises. *Science* 166: 1651.

Samuels, A., and P. Tyack. 2000. Flukeprints: A history of studying cetacean societies. In J. Mann, R. C. Connor, P. L. Tyack, and H. Whitehead, eds., *Cetacean Societies: Field Studies of Dolphins and Whales,* 9–44. Chicago: University of Chicago Press.

Tyack, P. 1991. Use of a telemetry device to identify which dolphin produces a sound. In K. Pryor and K. S. Norris, eds., *Dolphin Societies: Discoveries and Puzzles,* 319–44. Berkeley: University of California Press.

———. 1997. Development and social functions of signature whistles in bottlenose dolphins *Tursiops truncatus. Bioacoustics* 8: 21–46.

Whitehead, H., and J. Mann. 2000. Female reproductive strategies of cetaceans. In J. Mann, R. C. Connor, P. L. Tyack, and H. Whitehead, eds., *Cetacean Societies: Field Studies of Dolphins and Whales,* 219–46. Chicago: University of Chicago Press.

Woodcock, A. H., and A. F. McBride. 1951. Wave-riding dolphins. *Journal of Experimental Biology* 28: 215–17.

Würsig, B. 1986. Delphinid foraging strategies. In R. J. Schusterman, J. A. Thomas, and F. G. Wood, eds., *Dolphin Cognition and Behavior: A Comparative Approach*, 347–59. Hillsdale, N.J.: Lawrence Erlbaum.

Chapter 9

Abelson, R., and K. Nielsen. 1967. Ethics, history of. In P. Edwards, ed., *The Encyclopedia of Philosophy*, 3:70. New York: Macmillan and Free Press.

Bentham, J. 1976. A utilitarian view. In T. Regan and P. Singer, eds., *Animal Rights and Human Obligations*, 129–30. Englewood Cliffs, N.J.: Prentice Hall.

Budiansky, S. 1992. *Covenant of the Wild*. New York: William Morrow.

Cavalieri, P., and P. Singer, eds. 1993. *The Great Ape Project*. London: Fourth Estate.

Churcher, P. B., and J. B. Lawton. 1989. Beware of well-fed felines. *Natural History*, July, 40–47.

Cottingham, J. 1978. "A brute to the brutes?" Descartes' treatment of animals. *Philosophy* 53: 551–59.

———. 1986. *Descartes*. Oxford: Blackwell.

Evans, E. P. 1906. *The Criminal Prosecution and Capital Punishment of Animals*. New York: E. P. Dutton.

Fouts, R., and D. H. Fouts. 1993. Chimpanzees' use of sign language. In P. Cavalieri and P. Singer, eds., *The Great Ape Project: Equality beyond Humanity*, 28–41. New York: St. Martin's.

Gaita, R. 2002. *The Philosopher's Dog*. Melbourne: Text Publishing.

Goodall, J. 1988. *In the Shadow of Man*. Rev. ed. London: George Weidenfeld & Nicolson.

Hartley, A. 2002. Me Frodo, you Jane. *The Spectator*, June 29.

Hatley, P. J. 2003. *Feral Cat Colonies in Florida: The Fur and Feathers Are Flying*. A report to the U.S. Fish and Wildlife Service, January 2003. Gainesville, Fla.: University of Florida Conservation Clinic.

Miller, W. L. 1996. *Arguing about Slavery: The Great Battle in the United States Congress*. New York: Knopf.

Parker, J. 2000. Peter Singer's Animal Liberation. *AnimalRights.net*. <www.animalrights.net/print/articles/2000/000058.html>.

Regan, T., and P. Singer, eds. 1976. *Animal Rights and Human Obligations*. Englewood Cliffs, N.J.: Prentice Hall.

Rosenfield, L. C. 1969. *From Beast-Machine to Man-Machine*. New York: Octagon Books.

Singer, P. 1990. *Animal Liberation*. 2d ed. New York: Avon Books.

Wise, S. M. 2001. *Rattling the Cage: Toward Legal Rights for Animals*. New York: Perseus Books.

Acknowledgments

One of the pleasures of reaching the end of a project such as this is the opportunity to thank the colleagues and friends who did so much to make the journey an enjoyable one. First, both in time and contribution, Sam Elworthy—editor extraordinaire. Thank you, Sam, for your tireless enthusiasm. Without you as pilot, who knows where this ship might have ended up? Also providing guidance from the start: Bill Cannon. Thanks, Bill; I owe you, mate.

Several people commented on individual chapters. My thanks to Barry Ache, Jane Brockman, Bill Cannon, Graeme Cumming, Don Dewsbury, Mark Randell, and Herbert Terrace for their time and care. Likewise to Ray Anker, Stephen Budiansky, Bernard Wynne and an anonymous reviewer, all of whom read an earlier draft of the whole book and made very helpful comments.

Some years ago my friend John Staddon, in the preface to one of his books, thanked his employer, Duke University, for "a gracious environment conducive to scholarship." I confronted him about this: Surely it was very weak praise to say that a university offered conditions "conducive to scholarship"—what else could a university be for? I realize now that the scholarly environment I enjoy here at the University of Florida is not something that can be taken for granted on every university campus. Therefore it is with deep appreciation that I thank my colleagues in the Department of Psychology, especially our chairman, Martin Heesacker, and the dean of the College of Liberal Arts and Sciences, Neil Sullivan, for this gracious and supportive environment.

And to my wife, Ros, and son, Sam: thank you for making it all worthwhile.

Index